ESEA
TITLE-II GRANT

Green Algae, I:
Molecular Biology

Papers by
F. Wanka, Stephen H. Howell, Klaus Apel,
et al.

IN COOPERATION WITH THE
SMITHSONIAN SCIENCE INFORMATION EXCHANGE

Summaries of current research projects are included in the final section of this volume. Previously unpublished, these summaries were obtained from a search conducted by the Smithsonian Science Information Exchange, a national collection of information on ongoing and recently terminated research.

MSS Information Corporation
655 Madison Avenue, New York, N.Y. 10021

Library of Congress Cataloging in Publication Data

Wanka, F   comp.
    Green algae.

    (MSS topics in ecology series)
    CONTENTS: v. 1. Molecular biology.
    1. Chlorophyceae. I. Howell, Stephen H.,
joint comp. II. Apel, Klaus, joint comp.
III. Title.
QK569.C6W36   589'.47   73-10108
ISBN 0-8422-7134-1

TABLE OF CONTENTS

CREDITS AND ACKNOWLEDGEMENTS

Apel, Klaus; and Hans-Georg Schweiger, "Nuclear Dependency of Chloroplast Proteins in *Acetabularia*," *European Journal of Biochemistry*, 1972, 25:229-238.

Boynton, J.E.; N.W. Gillham; and J.F. Chabot, "Chloroplast Ribosome Deficient Mutants in the Green Alga *Chlamydomonas reinhardi* and the Question of Chloroplast Ribosome Function," *Journal of Cell Science*, 1972, 10:267-305.

Cattolico, Rose Ann; and Raymond F. Jones, "Isolation of Stable Ribosomal RNA from Whole Cells of *Chlamydomonas reinhardtii*," *Biochimica et Biophysica Acta*, 1972, 269:259-264.

Goodenough, Ursula W., "The Effects of Inhibitors of RNA and Protein Synthesis on Chloroplast Structure and Function in Wild-Type *Chlamydomonas reinhardi*," *The Journal of Cell Biology*, 1971, 50:35-49.

Goodenough, Ursula W.; and R.P. Levine, "The Effects of Inhibitors of RNA and Protein Synthesis on the Recovery of Chloroplast Ribosomes, Membrane Organization, and Photosynthetic Electron Transport in the *ac-20* Strain of *Chlamydomonas reinhardi*," *The Journal of Cell Biology*, 1971, 50:50-62.

Hoober, J. Kenneth, "A Major Polypeptide of Chloroplast Membranes of *Chlamydomonas reinhardi:* Evidence for Synthesis in the Cytoplasm as a Soluble Component," *The Journal of Cell Biology*, 1972, 52:84-96.

Howell, Stephen H.; and Linda L. Walker, "Synthesis of DNA in Toluene-Treated *Chlamydomonas reinhardi*," *Proceedings of the National Academy of Sciences*, 1972, 69:490-494.

Meins, Jr., Frederick; and Marc L. Abrams, "How Methionine and Glutamine Prevent Inhibition of Growth by Methionine Sulfoximine," *Biochimica et Biophysica Acta*, 1972, 266:307-311.

Wanka, F.; and J. Geraedts, "Effect of Temperature in the Regulation of DNA Synthesis in Synchronous Cultures of *Chlorella*," *Experimental Cell Research*, 1972, 71:188-192.

Wanka, F.; J. Moors; and F.N.C.M. Krijzer, "Dissociation of Nuclear DNA Replication from Concomitant Protein Synthesis in Synchronous Cultures of *Chlorella*," *Biochimica et Biophysica Acta*, 1972, 269:153-161.

Wanka, F.; and P.J.A. Schrauwen, "Selective Inhibition by Cycloheximide of Ribosomal RNA Synthesis in *Chlorella*," *Biochimica et Biophysica Acta*, 1971, 254:237-240.

PREFACE

This collection of current research on the green algae, a volume in the MSS Topics in Ecology Series, focuses on the macromolecular structures of these simple eukaryotes and their genomic control systems. The papers, all published in the period 1970-1972, are intended to bring the reader an up-to-date overview of the most recent research in the nucleic acids of green algae, the interrelationships between the nuclear and chloroplast genomes, and the synthesis and function of tubulins and other proteins.

A companion volume, *Green Algae II: Cytology*, presents papers dealing with the cytology of cell wall and reproductive forms as well as those dealing with the effects of inhibitors of macromolecular synthesis and of radiation. Current research on culture methods and taxonomic relations within this vast group is also presented.

DNA Synthesis in Green Algae and
Characterization of Chloroplast and
Nuclear DNA

EFFECT OF TEMPERATURE IN THE REGULATION OF DNA SYNTHESIS IN SYNCHRONOUS CULTURES OF *CHLORELLA*

F. WANKA and J. GERAEDTS

Synchronous cell cultures may be profitably used to study regulatory processes such as DNA replication during the cell cycle [1, 2]. These studies have revealed that the cell development of *Chlorella* and related green algae differs fundamentally from that of other eucaryotes. In cells of *Chlorella pyrenoidosa* 211/8b, which are synchronized by alternating 16 h light with 8 h dark, nuclear DNA replication ensues about 10 h after the beginning of the light period and usually comprises four coherent replication cycles, in which DNA duplications alternate with nuclear divisions. After four cycles are completed about 18 h after the beginning of the light period, spore formation occurs and the 16 daughter cells are released by the end of the dark period.

The number of successive replication cycles occurring between the 10th and 18th hour can be made less than the normal four by modifying the conditions of culture, e.g. light and temperature [1, 3]. Results obtained upon appropriate treatments have led to the proposal of a regulation model involving initiation of replication cycles by specific metabolites. According to the model, the concentration of a hypothetical precursor necessary for initiation of replication would increase in response to light-promoted processes [4]. In the dark, concentration of such precursor would fall to an ineffective level within $2\frac{1}{2}$ h or less, depending on the duration of the preceding light period [3]. The understanding of processes by which initiation of these cycles is affected is incomplete, primarily because no information is available on the chemical nature of this precursor. Recent results suggest that exhaustion of precursor in the dark might be related to protein synthesis [5]. This relationship was further investigated in the present study.

MATERIALS AND METHODS

Cell growth and experimental design

The *Chlorella* strain 211/8b from the algae collection of the University of Göttingen was used for the investigation. The cells were grown photo-autotrophically in suspension culture at 30°C, and synchrony was maintained by alternating periods of 16 h light and 8 h dark. The cultures were diluted 16 times at the end of each dark period. Further details have already been reported [6].

In order to avoid any possible complicating side effects caused by light, experiments were started 13 to 14 h after the beginning of the light period and were generally continued in uninterrupted darkness. Temperature fluctuation in the culture tubes were obviated by immersing in a water bath. In the cell suspension the desired temperature was acquired within 5 min. Cycloheximide (Actidione from Koch-Light, Ltd) was added to the cell suspension from an aqueous solution concentrated 100 fold. Cell numbers were determined with a Coulter Counter.

11

DNA determination

Samples of 50 ml were withdrawn from the cultures in duplicate and centrifuged for 2 min at 2 000 g. The cell sediment was preextracted twice for 30 min at room temperature with 0.2 N perchloric acid in 50 % ethanol and twice for 10 min at 60°C with ethanol–ether (3:1) in order to remove acid- and lipid-soluble impurities [7]. The nucleic acids were then extracted from the residue by heating for 6 h to 45°C in 0.5 N perchloric acid. The DNA content of the extracts was determined by the diphenylamine test according to Burton [8]. The average difference between determinations of duplicate samples was less than 2 %.

RESULTS

When *Chlorella* cultures were transferred to the dark 14 h after the beginning of the light period and exposed to various temperatures for 4 h, the resulting changes in final DNA content and cell number/ml were similar. For example temporary exposure to 15 and 16°C

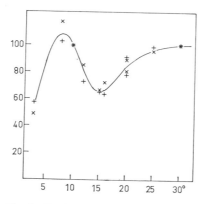

Fig. 1. Abscissa: temperature; *ordinate:* %.

Effect of a temporary lowering of temperature on the final DNA content and cell counts. A *Chlorella* culture was withdrawn from the light 14 h after the beginning of the light period, and divided into portions of 250 ml. These were grown for 4 h at various temperatures and subsequently brought back to 30°C. DNA content (×) and cell counts (+) were determined 20 h later. The values of the 30°C control cultures were set equal to 100 %.

changed the cell development to the point that final DNA contents and cell counts were only 60 to 70 % of the values in controls kept at 30°C (fig. 1). However, after a treatment at 8 or 10°C, the final DNA contents and cell counts were just as high or even slightly higher than in the 30°C controls. The low values found after a 3°C treatment will not be further considered, since it is known that cell activities other than DNA synthesis are seriously impaired at this low temperature [9]. The final DNA contents per cell varied between 1 and 1.2×10^{-13} g, in agreement with previously reported values [6, 10].

The rates of DNA synthesis decreased markedly soon after changing to low temperatures (fig. 2). In more detailed experiments, we found that during the first 2 h at 15°C the rate was only about 17 % of that in the control culture and at 10°C, about 11 %. Therefore a replication cycle at 15°C would be expected to require about 10 h (6 times as long as at 30°C), and about 15 h at 10°C. When the cultures were transferred back to 30°C after 3 h at a lower temperature, the DNA synthesis was immediately restored to about the same rate as before the treatment (fig. 2). That is, after bringing cells treated at 8 or 10°C back to 30°C, DNA synthesis continued at a normal rate until the DNA content approximated that of the untreated control and stopped as is characteristic of normal cultures in the dark. However, transferring cultures back from 15° to 30°C resulted in cessation of DNA synthesis when its content had increased only 60 %. Apparently, the cells completed the replication cycle which was in progress at the time of changing to 15°C, but only a small fraction of them were able to enter another cycle after that. In contrast with this, the DNA in cells remaining at 8 or 10°C increased as much as 150 %, indicating that most of the cells had preserved the capacity to enter an additional S phase and mitosis after transfer back to 30°C. Therefore, the final DNA content and cell counts evidently depend, to a large extent, on the degree to which the ability to undergo a further replication cycle is preserved during temperature treatment. This ability was also gradually lost when a 10°C treatment was extended beyond the usual 3 to 4 h.

It has been hypothesized that the loss of the ability to enter a new S phase might be caused by the depletion of specific metabolites in the dark [3], and recent results suggest that this process might be linked to protein

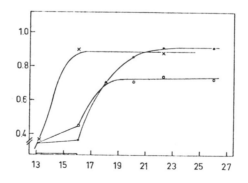

Figs 2, 3. Abscissa: time after beginning of the light period (hours); *ordinate:* DNA (μg/ml).
Fig. 2. Effect of a temporary lowering of temperature on the DNA synthesis. A *Chlorella* culture was withdrawn from the light 13 h after the beginning of the light period and divided into 3 portions. Two were grown at temperatures of 8 (●) and 15°C (○) for 3 h and then brought back to 30°C. DNA contents were determined at the times indicated. (×) 30°C control.

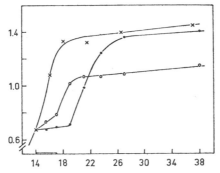

Fig. 3. Combined effect of a change in temperature and cycloheximide on DNA synthesis. A *Chlorella* culture was divided into 3 portions 14 h after the beginning of the light period. Two portions were transferred to a temperature of 15°C and cycloheximide was added to one of them. The inhibitor was removed 3 h later by centrifuging for 2 min at 1 000 g and resuspending the cell pellet in fresh nutrient solution. The cultures were further grown at 30°C and DNA was determined at the times indicated. ○, 15°C; ●, 15°C and 15 μM cycloheximide; ×, control.

synthesis [5]. It should be possible, therefore, to maintain a high level of such hypothetical precursors by inhibiting protein synthesis. Fig. 3 illustrates increase in DNA content in matched cultures at 15°C for 3 h, in the presence and absence, respectively, of 15 μM cycloheximide. Cycloheximide is known to inhibit synthesis of enzymes and bulk protein in *Chlorella* [5, 10, 11, 12]; and since nuclear DNA synthesis is strongly dependent on concomitant protein synthesis, DNA production is also blocked in the presence of cycloheximide [5, 13]. After removing the inhibitor and raising the temperature to 30°C the rate of DNA synthesis remained at a low level for about 2 h, and then returned to approxi-

15

mately a normal rate until the same DNA content was attained as in the control (fig. 3). Again in the 15°C-culture without cycloheximide, the final DNA content remained about 20 % lower than at 30°C.

In view of the possible involvement of protein synthesis, experiments on the temperature dependence of protein synthesis were performed which showed that the synthetic rate at 15 and 10°C was about 50 and 25 % of that of the control at 30°C. Thus, the effect of temperature on protein formation is less pronounced than the effect on DNA replication.

DISCUSSION

An outstanding feature of the synchronous *Chlorella* system is the possibility of controlling the regulation of DNA synthesis experimentally. Treatments affecting synthesis may be started when most cells have passed through the second S phase of a cell cycle, or after about 13 h in light. At this time, the general requirement for metabolites and energy is met by degradation of cellular starch [11] and consequently, replication is not retarded when cultures are placed in the dark. The only effects observed occur after about 2 h in the dark and consist of loosing the potential to initiate further replication cycles, which is concomitant with a premature transition to the subsequent developmental stage [1]. Synchronous cultures placed in the dark after 13 h of light undergo an average of three replication cycles instead of the normal four.

The present findings further substantiate the regulation model proposed earlier [3, 4]. According to this model a new S phase can be entered soon after completion of the preceding replication cycle. The initiation process would depend on the availability of specific metabolites which undergo depletion in the dark. An ineffective level of these metabolites is apparently reached within 2 h in the dark at 30°C, 3 h at 18°C, 4 h at 15°C and a much longer period at 10°C or less. The frequency with which initiation of replication cycles occur is also strongly reduced at lower temperatures because of the lengthening of the individual replication cycles; at 15°C for example this frequency is about one-sixth of that observed at 30°C. Thus, only 30 to 40 % of the cells will be able to complete the current replication cycle and enter the subsequent S phase before the exhaustion of the required precursors occurs, whereas at 30°C, all cells undergo an additional S phase and mitosis. Once the precursors are exhausted, the remaining 60 to 70 % of the cells do not undergo a further replication even if the temperature is changed back to 30°C. Such an exhaustion would also explain the failure to enter a new S phase after a temporary interruption of the DNA synthesis by hydroxyurea [5].

It has been suggested not only that the hypothetical precursor is consumed by the process of initiation itself, that is by becoming a component of some unknown initiation factor, but also that to a marked extent it might be made unavailable for initiation via a different reaction [3]. It seems now that the latter reaction is linked to protein synthesis and that the temperature effects discussed above might possibly be mediated via the different response of protein and DNA synthetic rates on temperature. A competition between protein formation and the initiation of DNA replication might also account for the observed inverse relationship between response of protein content and DNA content together with cell counts, to different wavelengths of light [14]. Tamiya, in discussing work of his own and of his co-workers on *Chlorella ellipsoidea* [15], suggested a competition between protein synthesis and the increase of inducer substances for cell division. Inducer substances might be sulfur-containing peptide-nucleotides appearing at a specific time in the cell cycle [16–18]. Similar compounds have been reported to stimulate cell division in *Chlorella* [19–21]. The relationship between such inducer substances and the precursor discussed above has still to be investigated.

We thank Mr J. Eygensteyn for technical assistance and Dr L. Douglas for kindly reviewing the English text.

REFERENCES

1. Wanka, F & Mulders, P F M, Arch Mikrobiol 58 (1967) 257.
2. Wanka, F, Joosten, H F P & De Grip, W J, Arch Mikrobiol 75 (1970) 25.
3. Wanka, F, Planta 79 (1968) 65.
4. — Arch Mikrobiol 34 (1959) 161.
5. Wanka, F, Moors, J & Krijzer, F W J M, Biochim biophys acta. Submitted for publication. 269,153
6. Wanka, F, Arch Mikrobiol 52 (1965) 305. (1972)
7. — Planta 58 (1962) 594.
8. Burton, K, Biochem j 62 (1956) 315.
9. Lorenzen, H, Flora 153 (1963) 554.
10. Wanka, F & Poels, C L M, Europ j biochem 9 (1969) 478.
11. Wanka, F, Joppen, M M J & Kuyper, Ch M A, Z Pflanzenphysiol 62 (1970) 146.

12. Schönherr, O Th & Wanka, F, Biochim biophys acta 232 (1971) 83.
13. Wanka, F & Moors, J, Biochem biophys res commun 41 (1970) 85.
14. Pirson, A & Kowallik, W, Photochem photobiol 3 (1964) 489.
15. Tamiya, H, J cell comp physiol 62, suppl. 1 (1963) 157.
16. Hase, E, Mihara, S & Tamiya, H, Biochim biophys acta 39 (1960) 381.
17. Hase, E, Mihara, S & Tamiya, H, Plant cell physiol 1 (1960) 131.
18. — Ibid 2 (1961) 9.
19. Vraná, D & Fencl, F, Folia microbiol 12 (1967) 432.
20. Goryunova, S V, Pusheva, M A & Gerasimenko, L M, Dokl akad nauk SSSR, (transl by Consultants Bureau, New York) 190 (1970) 63.
21. — Ibid 190 (1970) 69.

DISSOCIATION OF NUCLEAR DNA REPLICATION FROM CONCOMITANT PROTEIN SYNTHESIS IN SYNCHRONOUS CULTURES OF CHLORELLA

F. WANKA, J. MOORS AND F. N. C. M. KRIJZER

SUMMARY

The interrelationship between nuclear DNA replication and cytoplasmic protein synthesis in *Chlorella* was studied by use of hydroxyurea and cycloheximide as specific inhibitors.

15 μM cycloheximide completely blocked the synthesis of protein and depressed DNA and RNA formation by about 90 %. Protein synthesis resumed immediately after removal of the inhibitor, while DNA synthesis was fully restored after a lag time of about 2 h. 10 mM hydroxyurea completely inhibited DNA synthesis. A decrease in protein and RNA synthesis gradually ensued after about 3 h of exposure. DNA synthesis was restored immediately after removal of the inhibitor, but the final DNA levels were markedly lower than in control cultures. During the recovery from the hydroxyurea treatment about 50 % increase of the DNA content was observed in the presence of cycloheximide.

The results suggest that simultaneous synthesis of specific proteins required for DNA replication occurs during the S phase. These proteins continue to be synthesized when nuclear DNA replication is stopped by specific inhibitors, and can then be used to support subsequent DNA replication in the absence of protein synthesis.

INTRODUCTION

The cell cycles of *Chlorella* and related unicellular green algae differ fundamentally from those of most other eucaryotic cells. When *Chlorella* cultures are synchronized by alternating periods of 16 h light and 8 h dark, young cells undergo growth processes during the first 10 h after the beginning of the light period. Between the 10th and 18th h four cycles of DNA replication alternating with nuclear division usually take place, and the cell content is cleaved into 16 nucleate parts[1]. Within the next 4 h these naked protoplasts are furnished with cell walls and finally released from the mother cells. The number of replication cycles, and corresponding final cell numbers, are strongly influenced by minor changes of environmental conditions during the replication period.

20

Recently, a selective inhibition of nuclear DNA replication in *Chlorella pyrenoidosa* and *Saccharomyces cerevisiae* was observed when protein synthesis was inhibited by cycloheximide or by amino acid starvation, while syntheses of mitochondrial and chloroplast DNA continued[2,3]. However, the activities of several enzymes involved in DNA biosynthesis did not decrease significantly after exposure of *Chlorella* cells to cycloheximide for several hours, and *in vitro* activity of DNA polymerase was not affected by the inhibitor at a concentration of 600 μM[4,5]. It was suggested, therefore, that nuclear DNA replication requires concomitant formation of specific proteins during the S phase[2]. We report here some experiments on the uncoupling of DNA synthesis from such protein formation by use of specific inhibitors.

MATERIALS AND METHODS

Cell growth

The *Chlorella pyrenoidosa* strain 211/8b from the algae collection of the University of Göttingen was used for the experiments. The cells were grown photoautotrophically at 30° as suspension cultures[6]. Cell synchrony was induced by alternating periods of 16 h light and 8 h dark. The cultures were diluted to 1/16th at the end of each dark period. Inhibition experiments were usually started 13–14 h after the beginning of the light period and were performed in the dark in order to exclude complicating effects which might arise from interference with photosynthesis. Cycloheximide (Actidiono, Koch-Light Lab. Ltd.) and hydroxyurea (Schuchardt) were added to the cultures from 100-fold concentrated aqueous solutions. Cell numbers were determined with the aid of a Coulter Counter.

Chemical analyses

For determinations of cell composition 50-ml samples of cell suspension were centrifuged for 2 min at 2000 $\times g$ and the resulting pellet was used for analyses. Acid- and lipid-soluble impurities were removed by 2 extractions for 30 min with 0.2 M $HClO_4$ in 50 % ethanol at room temperature and, subsequently 3 extractions for 10 min with ethanol–ether (3 : 1, v/v) at 60°. The colorless sediment was washed with ice-cold 0.5 M $HClO_4$ and the nucleic acids were extracted by hydrolysis for 6 h in 0.5 M $HClO_4$ at 45° (ref. 7). DNA determination was carried out by the diphenylamine test according to the method of Burton[8]. The RNA content was calculated from the absorbance of the extracts at 260 nm by subtraction of the portion of absorption due to DNA. Protein was extracted from the cell residue with hot 1 M NaOH[7] and determined by the method of Lowry *et al.*[9] adapted for the high alkali content of the extract.

Labeling and preparation of DNA

[2-14C]Uracil (spec. act. 31 mCi/mmole; New England Nuclear Corp.) was used as a label, because thymidine is not incorporated in *Chlorella* DNA. Nucleic acids were extracted and DNA was carefully purified according to the procedure of Wanka *et al.*[10] After ribonuclease digestion the DNA was dissolved in 10 ml of 0.4 M NaCl containing 0.05 M phosphate buffer (pH 6.7) and adsorbed onto a column containing 5 g of kieselguhr coated with methylated albumin[11]. The column was washed with

20 ml of the same saline solution and subsequently eluted by 0.8 M NaCl in phosphate buffer (pH 6.7). The DNA was precipitated by addition of 2 vol. 96 % ethanol containing 1 % potassium acetate and redissolved in 0.02 M Tris–HCl buffer (pH 8).

Isopycnic centrifugation in a CsCl density gradient

The volume of the DNA solution was brought to 4.4 ml, and 5.5 g CsCl (Merck, p.a.) were added. The sample was centrifuged for 65 h in a Spinco L 2/65 B at 4° and 35 000 rev./min using an angle rotor Type 65 (ref. 12). Fractions of 8 drops were collected from the centrifuge tube and brought to 2 ml with 0.02 M Tris buffer (pH 8.5) and the absorbance was measured at 260 nm. The radioactivity of 1-ml aliquots was determined in a Packard Tri-Carb Liquid Scintillation Spectrometer, Model 3375, using a dioxane-based scintillation fluid[10].

RESULTS

Concentration dependence of the inhibiting effects of hydroxyurea and cycloheximide

Hydroxyurea is a strong inhibitor of DNA synthesis in *Chlorella*[13]. Results obtained with other organisms indicate that this effect might be due to a specific inhibition of the ribonucleoside diphosphate reduction[14, 15]. DNA synthesis was inhibited completely in the presence of hydroxyurea at concentrations exceeding 8 mM (Fig. 1). The concentration dependence was very similar to that found for ribonucleoside diphosphate reduction in mammalian cells *in vivo*[15] and in crude extracts of various organisms *in vitro*[16]. Protein formation was found to be reduced by only 30 % after 4.5 h of exposure to 16 mM hydroxyurea (Fig. 1) but very little inhibition was observed during the first 3 h of treatment (Fig. 4). This effect is compatible with the mild influence of hydroxyurea, particularly during short treatments, on

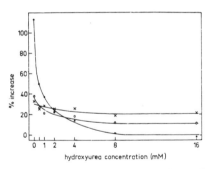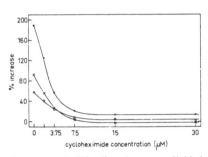

Fig. 1. Concentration dependence of the effect of hydroxyurea. A *Chlorella* culture was divided into 7 parts of 250 ml 13.5 h after the beginning of the light period and different concentrations of hydroxyurea were added as indicated. The cultures were placed in the dark and DNA (●–●), RNA (○–○) and protein (×–×) contents were determined after 4.5 h. The increase is expressed in per cent of the initial content.

Fig. 2. Concentration dependence of the effect of cycloheximide. A *Chlorella* culture was divided into 6 parts of 250 ml 13 h after the beginning of the light period and different concentrations of cycloheximide were added as indicated. The cultures were placed in the dark and DNA (●–●), RNA (○–○) and protein (×–×) contents were determined after 5 h. The increase is expressed in per cent of the initial content.

the increase of cellular DNA polymerase activity[13]. Therefore, the primary target of hydroxyurea in *Chlorella*, like in other organisms, seems to be the inhibition of nucleotide reductase. DNA synthesis is consequently prevented, and, upon increasing the exposure times, protein formation declines. The same sequence of events seem to explain the decline in RNA formation caused by hydroxyurea (Fig. 1 and unreported results). Reversal of the hydroxyurea effect by addition of deoxynucleosides cannot be obtained because of the failure of *Chlorella* to incorporate externally applied deoxythymidine and deoxycytidine into its DNA.

Cycloheximide blocked protein accumulation completely at concentrations of 7.5 μM and above (Fig. 2). 3 μM cycloheximide caused a 50 % inhibition of protein synthesis. This is very close to the concentration at which a 50 % inhibition of amino acid incorporation was reported in a cell free system of *Saccharomyces pastorianus*[17]. The concentration dependence of the inhibition of DNA and RNA synthesis was very similar, supporting the conclusion that the failure to replicate nuclear DNA—and, obviously, major RNA fractions also[18]—is due to the inhibition of cytoplasmic protein synthesis which, like in other eucaryotes[17, 19], appears to be the immediate target of the cycloheximide action in *Chlorella*. Inhibition of DNA synthesis was never complete; in fact, an increase of up to 10% was repeatedly observed at cycloheximide concentrations of 7.5 μM and above. This effect is due, at least in part, to uninhibited synthesis of satellite DNA[2].

Reversibility of the inhibitor effects

In general the effect of cycloheximide on DNA content was completely reversible if the exposure time was less than 6 h. Longer treatment resulted in significant numbers of non-viable cells. A DNA increase of more than 100 % was usually observed within 5 h after removal of the inhibitor if the experiment was started 13 h after the beginning of the light period. The final DNA contents were mostly equal

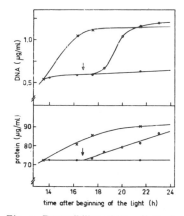

Fig. 3. Reversibility of the effect of cycloheximide. A *Chlorella* culture (1.1 · 10⁶ cells per ml) was exposed to 15 μM cycloheximide 13.5 h after the beginning of the light period and placed in the dark. The cells of part of the culture were collected by centrifugation 3 h later (arrow) and resuspended in an equal volume of nutrient solution. The change of the DNA and protein contents was determined with cycloheximide present continuously (●–●) or removed after 3 h (○–○) and in an untreated control culture (×–×).

to, or even greater than those of untreated control cultures (Fig. 3). However, the main recovery was always preceded by a lag time of about 2 h during which time the DNA increase was between 2 and 15 % only (4 experiments). Protein synthesis was resumed sooner but at the given accuracy of the protein determination a lag time of 1 h cannot be excluded. DNA contents, equal or greater than in the controls, were also obtained when cells were exposed to cycloheximide for 2 or 4 h at various times in the replication period (Table I). Corresponding numbers of daughter cells

TABLE I

RECOVERY FROM TREATMENT WITH CYCLOHEXIMIDE

To cultures of 300 ml 15 μM cycloheximide was added at different times of the replication period. The cells were collected 2 or 4 h later and resuspended in nutrient solution. All cultures obtained a light period of 16 h and were subsequently left in continuous dark in order to prevent the induction of a new division cycle. DNA, protein and cell numbers were estimated 40 h after the beginning of the light period.

Time of inhibition*	Protein ($\mu g/ml$)	RNA ($\mu g/ml$)	DNA ($\mu g/ml$)	Cells per ml $\times 10^{-6}$	g DNA per cell $\times 10^{13}$
10–12	60	13.6	1.08	10.3	1.05
10–14	58	12.4	1.14	10.5	1.09
12–14	63	14.7	1.19	12.0	1.00
12–16	61	12.5	1.34	12.4	1.08
14–16	62	14.8	1.23	12.0	1.02
14–18	56	11.7	1.15	11.5	1.00
16–18	66	14.7	1.05	10.5	1.00
16–20	64	13.9	1.12	11.2	1.00
Uninhibited control	67	15.1	0.99	10.0	0.99

* Hours after the beginning of the light period.

were obtained with normal DNA contents of about 10^{-13} g per cell[1], indicating that the DNA formed during the recovery has biological function. Smaller numbers of cells with a greater DNA content were obtained only when the cycloheximide treatment was towards the end of the replication period. Microscopical examination revealed that in this case a few large cells which failed to divide were responsible for the deviation. Protein and RNA contents were usually slightly less after a temporary exposure to cycloheximide (Table I).

When the reversibility of the hydroxyurea effect was studied the results were different with regard to two essential points (Fig. 4): (1) DNA synthesis was resumed at the normal rate immediately after removal of the inhibitor and (2) the final DNA content became less with increasing duration of the treatment, and was always significantly less than in control cultures. But an increase of the DNA content of at least 40 % took place, even after more than 4 h of exposure, provided that the experiment was started at or before the middle of the replication period. Such an amount of DNA formation is apparently required in order to complete the S phases which were disrupted by addition of the inhibitor. Proof that the effect of hydroxyurea is not due to a permanent impairment of the replication mechanism was sought by an experiment in which hydroxyurea and cycloheximide were present simultaneously. In such cultures the final DNA content was always significantly greater than in cultures

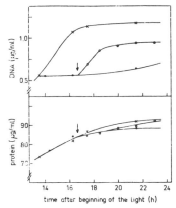

time after beginning of the light (h)

Fig. 4. Reversibility of the effect of hydroxyurea. The experiment was carried out as described in Fig. 3 except that 10 mM hydroxyurea was used as inhibitor instead of cycloheximide. Same symbols as in Fig. 3.

exposed to hydroxyurea only (Table II), suggesting that the limited recovery is due, at least in part, to some particular change in metabolic processes, which can be altered or prevented by cycloheximide. Such an alteration is supported by the finding of a similar lag time before the restoration of DNA synthesis as after exposure to cycloheximide only (Table II and Fig. 3).

TABLE II

COMBINED EFFECT OF HYDROXYUREA AND CYCLOHEXIMIDE

A *Chlorella* culture was divided into 4 parts of 250 ml 13.5 h after the beginning of the light period. Three parts were supplied with 15 μM cycloheximide or 10 mM hydroxyurea or a combination of both, and all were incubated in the dark. The cells were collected 3 h later and resuspended in nutrient solution. DNA contents were determined 2 and 7.5 h later, respectively.

Inhibitor	Time of sampling*	DNA ($\mu g/ml$)
None	13.5	0.35
Cycloheximide and hydroxyurea	18.5	0.35
Cycloheximide and hydroxyurea	24	0.69
Hydroxyurea	24	0.63
Cycloheximide	24	0.80
None	24	0.81

* Hours after the beginning of the light period.

However, the final DNA contents reached after combined action of hydroxyurea and cycloheximide were always lower than in cultures without hydroxyurea. Microscopical examination revealed that this is due to the fact that a few cells fail to divide and finally die. A similar cell degeneration in the presence of hydroxyurea has been observed with chicken fibroblasts *in vitro*[20] and might be caused by side effects, like alteration of DNA which has been observed at high concentrations of hydroxyurea *in vitro*[21]. Limited exposure to hydroxyurea had little effect on protein

25

and RNA synthesis (Fig. 4 and results not reported). If the hydroxyurea treatment was continued, a slow increase in DNA often ensued after 4–5 h. This may be ascribed to degradation of the inhibitor as suggested by Gauthier and Bassleer[20].

Uncoupling of DNA synthesis from protein synthesis

Having sufficient evidence that cycloheximide prevents DNA replication by inhibiting protein synthesis and that hydroxyurea does the same apparently by inhibiting nucleoside diphosphate reduction, it could now be asked whether it might be possible to uncouple nuclear DNA replication from simultaneous protein synthesis in the cytoplasm. To this end DNA synthesis in a *Chlorella* culture was inhibited by adding 10 mM hydroxyurea 13.5 h after the beginning of the light period. The inhibitor solution was replaced after 3 h by equal volumes of pure nutrient solution, or nutrient solution containing 15 μM cycloheximide. Different extents of DNA synthesis, ranging between 28 and 63 % (5 experiments), were found in the presence of cycloheximide. These values varied between 60 and 100 % compared to those found during recovery in the absence of cycloheximide. But in all experiments recovery from the hydroxyurea treatment in the presence of cycloheximide proceeded much more slowly than in pure nutrient solution. No significant changes of the protein contents were found during the 6- or 8-h exposure to cycloheximide, the final values being between 97 and 103 % of those present initially.

Proof that the DNA synthesized in the presence of cycloheximide was primarily nuclear was obtained in a final experiment. DNA replication was prevented for 2 h by addition of hydroxyurea. The cells were allowed to recover for 7 h with [14C]-

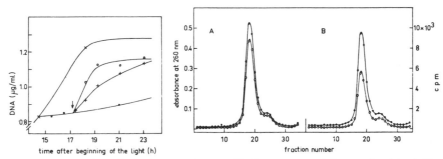

Fig. 5. Recovery from the hydroxyurea treatment in the presence and absence of cycloheximide A *Chlorella* culture (1.1 · 10⁶ cells per ml) was exposed to 10 mM hydroxyurea 14.5 h after the beginning of the light period and placed into the dark. The cells of the major part of the culture were collected by centrifugation after 2.75 h (arrow) and resuspended in an equal volume of nutrient solution. The sample was divided into 2 parts and 15 μM cycloheximide was added to one of them. Symbols are as follows: ●–●, continuous presence of hydroxyurea; ○–○, recovery in pure nutrient solution; +–+, recovery in the presence of 15 μM cycloheximide; ×–×, untreated control.

Fig. 6. Incorporation of labeled uracil into DNA during recovery from the hydroxyurea treatment. A *Chlorella* culture of 600 ml (1.6 · 10⁶ cells per ml) was placed into the dark and 5 mM hydroxyurea were added. The cells were collected after 2 h and resuspended in 200 ml of nutrient solution. The sample was divided into 2 parts. To one part was added 10 μCi of [14C]uracil and to the other, 10 μCi [14C]uracil *plus* 15 μM cycloheximide. Another 10 μCi of [14C]uracil were added to each culture 3.5 h later. The cells were collected 7 h after removal of the hydroxyurea and the DNA was extracted and banded in a preparative CsCl density gradient as described under Materials and Methods. ●–●, absorbance; ○–○, radioactivity. A, without cycloheximide; B, with cycloheximide present. Fraction 1 is bottom of the tube.

labeled uracil present in the growth medium. The purified DNA was banded in a preparative CsCl density gradient. Almost 90 % of the label was associated with the major *i.e.* nuclear DNA component (buoyant density 1.710 g/cm^3)[10]. A similar distribution of the radioactivity was found when protein synthesis was inhibited by cycloheximide during recovery from hydroxyurea treatment (Fig. 6), indicating that 80-90 % of the newly synthesized DNA was indeed nuclear.

DISCUSSION

The present and previous investigations show that both *Chlorella* and yeast fail to replicate nuclear DNA when cytoplasmic protein synthesis is prevented by specific inhibitors or by amino acid starvation[2,3]. Corresponding results with different organisms[22-25] suggest that, in general, nuclear DNA replication in eucaryotes is dependent on concomitant cytoplasmic protein synthesis. It is unlikely that the proteins in question are enzymes which are involved in DNA biosynthesis, if one considers that stable levels of such enzymes are usually observed when cells are exposed to cycloheximide, and that enzyme activities *in vitro* are not affected by the inhibitor[4,5,23,24]. The possibility must therefore be considered that nuclear DNA synthesis *in vivo* is intimately linked to simultaneous formation of certain histones[26,27] or other chromosomal proteins.

The results presented above demonstrate that it is possible to uncouple nuclear DNA replication from concomitant cytoplasmic protein synthesis. A considerable amount of nuclear DNA is formed in the presence of cycloheximide after temporarily blocking DNA synthesis by hydroxyurea. It has already been shown that under the growth conditions employed most nuclei are in some stage of the second or third S phase 13.5 h after the beginning of the light period[1]. When DNA synthesis is interrupted at this time by hydroxyurea for 2 or 3 h, the subsequent increase of the DNA content in the absence of cytoplasmic protein synthesis is usually about 50 %. One possible explanation might be that sufficient amounts of specific proteins accumulate to enable all nuclei to complete that particular S phase, but not to enter into the subsequent S phase. Beginning a new S phase seems to require additional initiation factors which cannot be formed unless the preceding S phase has been completed. Elucidation of the nature and function of the different kinds of protein factors involved will have to await future investigations. Histones are possible candidates, because of the restriction of their synthesis to the S phase. It has been shown that histone accumulation in *Physarum* continues when DNA replication is blocked by addition of fluorodeoxyuridine[28], although, in mammalian cells synthesis of certain histone fractions seems to require simultaneous DNA replication[26,29]. It is not yet clear in which way proteins might be responsible for the unexplained finding that the rate of DNA synthesis after the hydroxyurea treatment is much faster under conditions allowing protein synthesis, than in the presence of cycloheximide.

The *Chlorella* cells lose the capacity to undergo additional S phases when DNA synthesis in the dark is temporarily inhibited by hydroxyurea, but they fully preserve this capacity when cycloheximide is used as inhibitor alone and partly when it is used in combination with hydroxyurea. These effects are probably related to the processes which regulate the initiation of new replication cycles[30]. However, their implication is not yet fully understood.

27

ACKNOWLEDGEMENTS

We are very grateful to Mr J. Eygensteyn and to Mr J. M. A. Aelen for skilled technical assistance and to Dr R. D. MacElroy for kindly correcting the English typescript. The investigations were supported in part by the Netherlands Foundation for Chemical Research (S.O.N.) with financial aid of the Netherlands Organization for the Advancement of Pure Research (Z.W.O.).

REFERENCES

1 F. Wanka and P. F. M. Mulders, *Arch. Mikrobiol.*, 58 (1967) 257.
2 F. Wanka and J. Moors, *Biochem. Biophys. Res. Commun.*, 41 (1970) 85.
3 L. I. Grossman, E. S. Goldring and J. Marmur, *J. Mol. Biol.*, 46 (1969) 367.
4 F. Wanka and C. L. M. Poels, *Eur. J. Biochem.*, 9 (1969) 478.
5 O. Th. Schönherr and F. Wanka, *Biochim. Biophys. Acta*, 232 (1971) 83.
6 F. Wanka, *Arch. Mikrobiol.*, 52 (1965) 305.
7 F. Wanka, *Planta*, 58 (1962) 594.
8 K. Burton, *Biochem. J.*, 62 (1956) 315.
9 O. H. Lowry, N. J. Rosebrough, A. L. Farr and R. J. Randall, *J. Biol. Chem.*, 193 (1951) 265.
10 F. Wanka, H. F. P. Joosten and W. J. De Grip, *Arch. Mikrobiol.*, 75 (1970) 25.
11 J. Marmur, *J. Mol. Biol.*, 3 (1961) 208.
12 W. G. Flamm, H. E. Bond and H. E. Burr, *Biochim. Biophys. Acta*, 129 (1966) 310.
13 O. Th. Schönherr, Dissertation of the Faculty of Science, University of Nijmegen, 1969.
14 C. W. Young and S. Hodas, *Science*, 146 (1964) 1172.
15 L. Skoog and B. Nordenskjöld, *Eur. J. Biochem.*, 19 (1971) 81.
16 H. L. Elford, *Biochem. Biophys. Res. Commun.*, 33 (1968) 129.
17 M. R. Siegel and H. D. Sisler, *Biochim. Biophys. Acta*, 87 (1964) 83.
18 F. Wanka and P. J. A. Schrauwen, *Biochim. Biophys. Acta*, 254 (1971) 237.
19 W. McKeehan and B. Hardesty, *Biochem. Biophys. Res. Commun.*, 36 (1969) 625.
20 M. P. Gauthier and M. R. Bassleer, *C. R. Acad. Sci. Paris*, 272 (1971) 265.
21 S. J. Jacobs and H. S. Rosenkranz, *Cancer Res.*, 30 (1970) 1084.
22 J. E. Cummins and H. P. Rusch, *J. Cell Biol.*, 31 (1966) 577.
23 J. H. Kim, A. S. Gelbard and A. G. Perez, *Exp. Cell Res.*, 53 (1968) 478.
24 R. F. Brown, T. Umeda, S. Takai and I. Lieberman, *Biochim. Biophys. Acta*, 209 (1970) 49.
25 B. G. Weiss, *J. Cell Physiol.*, 73 (1969) 85.
26 A. Sadgopal and J. Bonner, *Biochim. Biophys. Acta*, 186 (1969) 349.
27 D. M. Prescott, *J. Cell Biol.*, 31 (1966) 1.
28 J. Mohberg and H. P. Rusch, *Arch. Biochem. Biophys.*, 138 (1970) 418.
29 D. Gallwitz and G. G. Mueller, *J. Biol. Chem.*, 244 (1969) 5947.
30 F. Wanka and J. Geraedts, *Exp. Cell Res.*, in the press.

Synthesis of DNA in Toluene-Treated
Chlamydomonas reinhardi

STEPHEN H. HOWELL AND LINDA L. WALKER

ABSTRACT   Toluene-treated *Chlamydomonas rein-hardi* incorporate deoxynucleoside triphosphates (dNTPs) into DNA. Incorporation requires all four dNTPs, but requires neither ATP nor added DNA. The incorporation reaction is linear for nearly 20 min. The product of synthesis in treated cells is mainly chloroplast (β-component) DNA. However, a small quantity of nuclear (α-component) DNA is also synthesized. Synchronously grown cells are most efficient in dNTP incorporation at a time in the cell cycle when chloroplast DNA is normally replicated. Toluene treatment disrupts the internal morphology of the cell, but leaves the outer membrane of the chloroplast intact.

In *Chlamydomonas*, nuclear and chloroplast DNA are independently replicated. The nuclear and chloroplast DNA, which can be distinguished by density differences, replicate semiconservatively at least once per generation (1, 2). The two DNA replicating systems appear to be independent of each other, in that chloroplast and nuclear DNA are synthesized at different times in the cell cycle (2).

In this paper, we have adapted the toluene treatment technique of Moses and Richardson (3) to *Chlamydomonas* in order to study the two DNA replicating systems. Toluene-treated *Chlamydomonas* cells are permeable to deoxynucleoside triphosphates (dNTPs) and incorporate them into DNA. Treated cells remain intact, but the integrity of most membrane-bound organelles is lost. Only the chloroplast retains its outer limiting membrane and, hence, its separateness from other cell components.

Toluene-treated cells incorporate dNTPs mainly into chloroplast DNA. These cells also incorporate a smaller amount into nuclear DNA. The capacity of toluene-treated cells to synthesize DNA varies throughout the cell cycle. Cells incorporate dNTPs most efficiently during the period of normal synthesis of chloroplast DNA. Therefore, DNA syn-

29

thesis observed in toluene-treated cells may reflect normal replication of chloroplast DNA.

MATERIALS AND METHODS

Chlamydomonas reinhardi strain 137C (mt^+), provided by Dr. W. B. Ebersold, was grown in high salt medium (HSM) (4, 5) at 27°C under 500 candelas of continuous light. Synchronous growth conditions at 21°C were described by Kates and Jones (6). Cells in logarithmic-phase growth (2–5 × 10^6 cells/ml) were harvested by centrifugation at 5000 × g for 10 min and washed with toluene-treatment buffer [10 mM Tris·HCl (pH 8.0)–1.0 mM dithiothreitol–0.5 mM EDTA–10 mM KCl]. Cells were resuspended to about 5 × 10^7 cells/ml in the same buffer. The suspension was made 0.5% in toluene and shaken on a rotator (about 60 inversions/min) for 15 min at room temperature. Treated cells were washed once in an equal volume of this buffer.

DNA was extracted from *Chlamydomonas* by a procedure modified from Chiang and Sueoka (2). Modifications included Pronase treatment [500 μg/ml in 1.0 M NaCl–0.015 M Na$_3$ citrate, (pH 7.0)] at 60°C for 2 hr, and phenol extraction. The extracted DNA was analyzed by equilibrium sedimentation according to Hotta and Stern (7). Enzymatic digestion of the labeled DNA was described by Sampson *et al.* (8), and the digestion products were analyzed by thin layer chromatography by the procedure of Randerath and Randerath (9).

Samples of cells for electron microscopy were fixed in 3% glutaraldehyde and post-fixed in 2% OsO$_4$. Standard embedding, staining, and sectioning procedures were used in subsequent preparation.

[^{3}H]TTP, [^{3}H]dATP, and [^{3}H]dCTP were products of New England Nuclear, pancreatic DNase was from Worthington, and BrdUTP was a gift from Dr. Douglas W. Smith.

RESULTS

Incorporation kinetics

The incorporation of [methyl-^{3}H]TTP by toluene-treated *Chlamydomonas* cells proceeds linearly at 30°C for nearly 20 min (Fig. 1). The incorporation rate was proportional to cell concentration to 1.0 × 10^8 cells/ml (2.0 × 10^7 cells/reaction assay). The extent of reaction was likewise proportional to cell concentration throughout that range. Freshly-treated cells incorporated [^{3}H]TTP most efficiently. Maximum rates varied from 1.5 to 3.0 pmol of TMP incorporated per min per 10^9 cells for continuously grown cells. Untreated cells failed to incorporate [^{3}H]TTP.

Reaction requirements

The incorporation of [³H]TTP requires all four dNTPs. Only 12% of the original activity was obtained in the absence of the other three dNTPs (Table 1). [³H]dCTP or [³H]dATP was incorporated as efficiently as [³H]TTP. It appeared that a breakdown product of [³H]TTP, such as the ³H-labeled deoxynucleoside, was not incorporated, since treated cells failed to incorporate [³H]dT in the presence or absence of the other dNTPs. No requirement for ATP was observed; in fact, ATP concentrations from 0.2 to 2.0 mM consistently reduced [³H]TTP incorporation by about 20%. Incorporation is not stimulated by the addition of either sonicated salmon-sperm DNA or poly(dAT).

Incorporation of [³H]TTP is inhibited by EDTA (Table 1). The optimum Mg^{++} concentration for the reaction was fairly high (25 mM); however, this requirement could be substituted by a combination of both 5 mM $MgCl_2$ and 25 mM KCl. Neither Mn^{++} or Ca^{++} effectively promoted the reaction in the absence of Mg^{++}.

Product analysis

The labeled product of [³H]TTP incorporation is stable in alkali, but can be solubilized by either 1 N HCl at 70°C or pancreatic DNase. The product of pancreatic DNase and venom-phosphodiesterase digestion cochromatographed with TMP. It was concluded that the product of [³H]TTP incorporation is DNA. The DNA labeled with [³H]TTP was more difficult to extract than bulk α- or β-component DNA. While bulk DNA is readily extracted by Sarkosyl–phenol (2), the extraction of DNA labeled with [³H]TTP required prolonged incubation with Pronase at 60°C before Sarkosyl–phenol extraction.

The ³H-labeled product was analyzed with neutral CsCl density gradients (Fig. 2). Although 85% of the DNA labeled with [¹⁴C]adenine is α-component or nuclear DNA (4), only 15% of the ³H-labeled product bands with a density similar to that of the α-component [ρ = 1.723 g/ml (3)]. The remaining product bands broadly at a density lighter than the nuclear component, but similar to that of the chloroplast DNA component (β-component). Two lighter peaks were consistently observed in the ³H-labeled product—one at ρ = 1.695 g/ml and the other at ρ = 1.702–1.705. [¹⁴C]Adenine-labeled β-component can often be resolved into two bands with nearly these same densities. Therefore, it appears that the product of [³H]TTP incorporation is predominantly β-component or chloroplast DNA.

31

Density-shift experiment

The product synthesized in the presence of the analog BrdUTP was analyzed to determine whether synthesis is semiconservative. Continuously-grown *Chlamydomonas* were harvested, treated with toluene, and incubated in the presence of [³H]dATP, dGTP, dCTP, and TTP or BrdUTP (Fig. 3). The [³H]dATP-labeled product synthesized in the absence of BrdUTP shows a density profile similar to the [³H]TTP-labeled product (α component, $\rho = 1.723$ g/ml, β components, $\rho = 1.695$ and 1.705 g/ml). Synthesis in the

TABLE 1. *Requirements for [³H]TTP incorporation by toluene-treated Chlamydomonas*

System	% activity
Complete	100
−dGTP	14
−dATP	20
−dCTP	22
−dGTP, dATP, dCTP	12
−MgCl₂	26
−KCl	66
−2-mercaptoethanol	54
+ATP	81
+EDTA	7
+ sonicated DNA	87
+ poly(dAT)	98

Contents of complete reaction as in Fig. 1. ATP was 0.2 mM, sonicated salmon-sperm DNA was 10 μg/ml, and poly(dAT) 27 μM, where indicated.

32

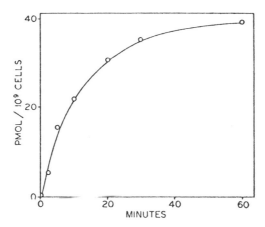

FIG. 1. Time course of incorporation of [³H]TTP by toluene-treated *Chlamydomonas*. 0.2-ml reaction mixtures contained 15 μmol of Tris·HCl (pH 7.5)–1 μmol MgCl₂–5 μmol KCl–0.2 μmol 2-mercaptoethanol–1 μmol [³H]TTP (2.0 Ci/mmol)–1 μmol (each) dCTP, dGTP, dATP, and 1.0 × 10⁷ cells. After incubation at 30°C, the reaction was stopped by the addition of 0.2 ml of cold 0.2 M sodium pyrophosphate and 0.5 ml of 10% trichloroacetic acid. Samples were filtered through Whatman GF/C filters (2.4 cm), then were washed five times with 5 ml of cold 10% acid and twice with 3 ml of 95% ethanol. The filters were dried and radioactivity was measured by liquid scintillation counting.

presence of BrdUTP appears to shift both major components to a greater density (α component, ρ = 1.728 g/ml, β components, ρ = 1.714 and 1.702 g/ml). The extent to which these components shift (α component $\Delta\rho$ = 5 mg/ml, β components $\Delta\rho$ = 8 mg/ml) is less than that expected for semi-conservative DNA synthesis. It was estimated that each component was displaced 15% in the direction of the fully-hybrid DNA molecule. [The expected density for the fully-substituted hybrid molecule was extrapolated in terms of

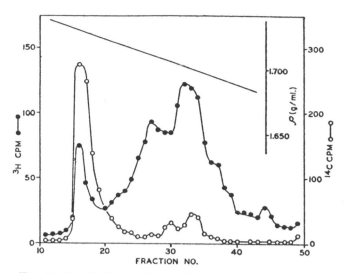

FIG. 2. Equilibrium sedimentation analysis of the product of [³H]TTP incorporation by toluene-treated cells. 30 ml of cells were grown in [¹⁴C]adenine to 5×10^8 cells/ml. Cells were harvested and treated with 0.5% toluene. [³H]TTP incorporation (6.1 Ci/mmol) was done as in Fig. 1 for 15 min at 30°C. The reaction was stopped by addition of 10 ml of cold 0.15 M NaCl–0.015 M Na₃ citrate. The cells were washed once with the same solvent, and DNA was extracted. The extracted DNA was centrifuged to equilibrium in neutral CsCl for 40 hr at 20°C at 35,000 rpm in a Beckman 50 Ti rotor. 60 fractions were collected, their refractive index was determined, carrier herring-sperm DNA was added, and the fractions were precipitated with 10% acid and filtered on Whatman GF/C filters.

G–C content from the density shift of 5-BrdU-substituted *Escherichia coli* TAU⁻ DNA, $\Delta\rho$ = 44 mg/ml, G–C content = 50% (10).] The limited density-shift observed, however, indicates that synthesis in the toluene-treated cells is not restricted to extremely short DNA segments characteristic of nonconservative or "repair" DNA synthesis (10).

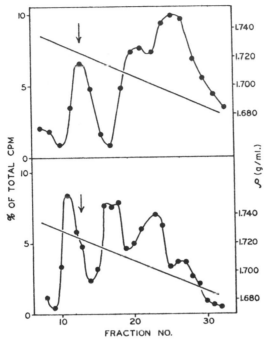

FIG. 3. Equilibrium sedimentation analysis of the product of [³H]dATP incorporation in the absence (*upper panel*) and presence (*lower panel*) of BrdUTP. Incorporation was as in Fig. 2, except that [³H]dATP (7.2 Ci/mmol) was used and BrdUTP was substituted for TTP in the *lower panel*. Fractions in the *upper panel* have been normalized to the same density as *lower panel* positions. Arrow indicates the density of ¹⁴C-labeled α-component marker ($\rho = 1.723$ g/ml).

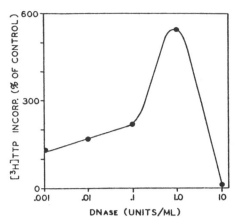

Fig. 4. Rate of [³H]TTP incorporation in reaction mixtures that contain various concentrations of pancreatic DNase. Reaction mixtures were the same as in Fig. 1, except that pancreatic DNase was added at 0 time in the reaction. The reactions were for 10 min at 30°C. The rate is expressed in terms of the percent of the control reaction (minus DNase).

Stimulation of synthesis by DNase

DNA synthesis in toluene-treated *Chlamydomonas* cells can be enhanced by the addition of pancreatic DNase to the reaction mixture. The same phenomenon has been reported for toluene-treated *E. coli* cells (3). In these experiments, DNA synthesis was stimulated more than 5-fold by the addition of 1.0 unit per ml of pancreatic DNase (Fig. 4). No incorporation into an acid-insoluble product was observed above 10 units per ml. Stimulation by DNase argues that toluene-treated cells can perform DNA repair synthesis. The reduced incorporation at higher DNase concentrations suggests that the stimulation effect is overbalanced by either the destruction of the newly formed product or the DNA template.

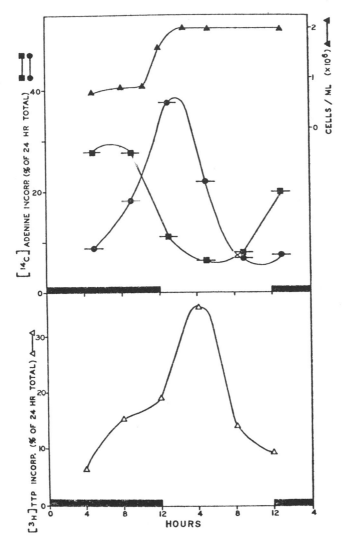

FIG. 5. Incorporation of [³H]TTP by toluenized cells, and incorporation of [¹⁴C]adenine during synchronous growth of cells. Cells were grown in a 12-hr dark, 12-hr light cycle, and analyzed at indicated intervals during a 24-hr period. In the *upper panel* cell number (cells/ml ▲——▲) was determined by hemocytometer counting. 2-hr incorporation of [¹⁴C]adenine into α-component DNA ■——■ (ρ = 1.713–1.733 g/ml) or into β-component DNA ●——● (ρ = 1.685–1.705 g/ml) is expressed as percent of total incorporation into that component during the 24-hr cycle. In the *lower panel* an aliquot of cells was harvested every 4 hr and treated with toluene. [³H]TTP incorporation into DNA at 30°C for 15 min (△——△) was determined for each sample and expressed in terms of percent of total incorporation in a 24-hr cycle.

37

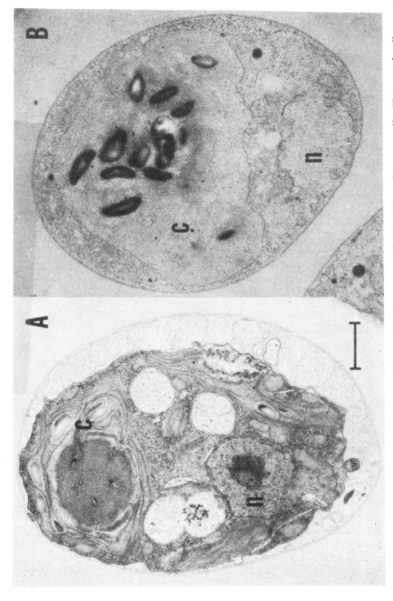

Fig. 6. Electron microphotographs of untreated (A) and toluene-treated (B) Chlamydomonas cells. Untreated cells were handled exactly as were treated, without the addition of toluene to buffer. Untreated cell appears somewhat plasmolyzed from the effects of buffer. Chloroplast (c) in treated cell remains as a unit, recognizable by the included starch granules. Nucleus (n) shows only vestiges in the toluene-treated cells. Bar = 1 μm.

The ^{3}H-labeled DNA is obviously accessible to the added DNase. It might seem that the reaction occurs extracellularly, or that the reaction products are exuded out from treated cells. However, the supernatant fluid from a reaction mixture centrifuged before or after incubation with [^{3}H]TTP contains no more than 10% of the acid-insoluble radioactivity.

Incorporation of [^{3}H]TTP during the cell cycle

Chlamydomonas cells can be synchronized by a 12-hr light–12-hr dark cycle. We tested for the efficiency of incorporation of [^{3}H]TTP into DNA at various times in a single cell cycle.

In these experiments (Fig. 5), cells had tripled in number at the end of the 12-hr dark period. Short-term (2-hr) exposure to [^{14}C]-adenine, and subsequent analysis of the labeled DNA on CsCl density gradients, defined the periods of α- and β-component DNA synthesis. The maximum rate of α-component synthesis preceded the formation of daughter cells by 2- to 4-hr, and β-component synthesis largely followed division by the same time interval. This timing of the replication of α- and β-components is consistent with the findings of Chiang and Sueoka (2).

Marked changes were seen in the efficiency of incorporation of [^{3}H]TTP into cells treated with toluene at various time intervals. Cells treated during the interval in which chloroplast β-component DNA is sythesized were 5-times more efficient in [^{3}H]TTP incorporation than at other times in the cell cyle—including the interval of α-component synthesis.

However, it is interesting that the proportion of α- or β-component synthesized in the toluene-treated cells did not vary considerably during the cell cycle. The density profiles of [^{3}H]TTP-labeled DNA are similar regardless of the time in the cell cycle when the cells were treated with toluene.

Ultrastructural analysis of the toluene-treated cells

Toluene-treated cells were observed in thin sections by electron microscopy. Such cells were distinctly more electron transparent than were untreated cells (Fig. 6). While the boundary of the cell remained intact, the inner cell membranes were in considerable disarray. One striking exception was the chloroplast outer membrane, which survived the treatment and retained such chloroplastic inclusions as starch granules. The chloroplast in the treated cell lost its typical cup-shaped appearance, but seemed to retain its general size. Only vestiges of the nucleus or nuclear membrane were seen.

DISCUSSION

Toluene-treated *Chlamydomonas* incorporate deoxynucleotide triphosphates into DNA. Nearly 85% of DNA synthesized in toluenized *Chlamydomonas* is β-component (chloroplast) DNA; only 15% is α-component (nuclear) DNA. The observed synthesis, then, is not proportional to the amount of each component in the cell. Chiang and Sueoka (3) have reported, and we have confirmed, that 85% of the cellular DNA is α-component and 15% is β-component. Therefore, it appears that chloroplast DNA is synthesized at greater rate than nuclear DNA in the toluenized cell.

In view of the preferential synthesis of chloroplast DNA, it is interesting to note that the outer chloroplast membrane remains intact in the toluenized cell, while other membranous organelles and the inner membranes of the chloroplast are disordered by the treatment. If the penetration of deoxynucleotides was the principal limitation on rate in synthesis by toluenized cells, it would seem more likely that the chloroplast would be least efficient in TTP incorporation, since the chloroplast outer membrane seems to present an additional barrier to penetration. It may be that the chloroplast system appears most efficient only because the nuclear system was damaged by toluene treatment; this assumption is likely, since only remnants of the orginal nuclear structure are seen in the treated cells.

The maximum rate of DNA synthesis with continuously grown *Chlamydomonas* cells (3.0 pmol per min per 10^9 cells) is only 2% of the rate reported by Moses and Richardson for toluenized *E. coli* cells. The *in vivo* rate of chloroplast DNA synthesis in continuously-grown *Chlamydomonas* has not been reported; however, a rough estimation of that rate can be made from the density-shift experiments of Sueoka (1). Sueoka found that 2.5 hr of growth at 25°C was required to convert 50% of the parental α-component DNA molecules to hybrid density. Therefore, a single round of DNA replication of the α-component would probably be completed in 5 hr. If one assumes the same time duration for the β-component synthesis that was observed during synchronous growth in these experiments, then the rate of synthesis of the β-component would be about 180 pmol per min per 10^9 cells. Again, the synthetic rate in toluenized *Chlamydomonas* is about 2% of the expected rate for the β-component *in vivo*. The rate-limiting step in the toluenized system is not known.

The replication of DNA during 15 min of linear synthesis is not fully semiconservative DNA synthesis. The density displacement is 15% of the displacement expected for complete hybrid-density molecules. The failure to achieve full semiconservative synthesis may be caused by the small amount of total DNA synthesis during the 15-min interval, by the synthesis of DNA from many growing points in the chloroplast genome, or by the dilution of density analog (BrdUTP) by internal TTP pools. However, since a limited density-shift is observed with all components during synthesis with Brd-UTP, synthesis is not restricted to extremely short segments of DNA characteristic of "repair-replication" DNA synthesis (10). DNA synthesis in toluenized *Chlamydomonas*, nonetheless, can be stimulated by DNase. This finding suggests that the treated cells can additionally respond to a repair function.

The synchronous-cell experiments reported here strongly argue that the DNA synthesis observed in the toluenized cell is a manifestation of the chloroplast DNA replicating system. The maximum rate of [³H]TTP incorporation is coincident with that period in the cell cycle when chloroplast DNA is normally replicated. The incorporation rate is reduced during the period of normal nuclear DNA synthesis. An alternative explanation to account for this observation is that the apparent rate changes might be due to a changed permeability or the effects of precursor pools in treated cells during this interval.

The toluenized cell system offers two new opportunities for the study of DNA synthesis in *Chlamydomonas*. First, it allows for the rapid screening of mutants defective in DNA synthesis. With a comparison of the rate of [³H]adenine incorporation into DNA in untreated cells and [³H]TTP incorporation into toluenized cells, one could easily distinguish between defects in the nuclear or chloroplast DNA-replicating systems. Second, the toluenized cells allow for the specific labeling of DNA by a thymidine derivative. *Chlamydomonas* will not efficiently incorporate [³H]thymidine from growth medium into cellular DNA (unpublished observations).

We thank Dr. D. W. Smith and Ms. D. M. Baumgartel for their helpful discussions. This research has been supported by the National Science Foundation (GB-30237) and the American Cancer Society, California Division (553).

1. Sueoka, N. (1960) *Proc. Nat. Acad. Sci. USA* **46**, 83–91.
2. Chiang, K.-S. & Sueoka, N. (1967) *Proc. Nat. Acad. Sci. USA* **57**, 1506–1513.
3. Moses, R. E. & Richardson, C. C. (1970) *Proc. Nat. Acad. Sci. USA* **67**, 674–681.
4. Sueoka, N., Chiang, K.-S. & Kates, J. R. (1967) *J. Mol. Biol.* **25**, 47–66.
5. Chiang, K.-S., Kates, J. R., Jones, R. F. & Sueoka, N. (1970) *Develop. Biol.* **22**, 655–669.
6. Kates, J. R. & Jones, R. F. (1964) *J. Cell Comp. Physiol.* **63**, 157–163.
7. Hotta, Y. & Stern, H. (1971) *J. Mol. Biol.* **55**, 337–355.
8. Sampson, M., Katoh, A., Hotta, Y. & Stern, H. (1963)

Proc. Nat. Acad. Sci. USA **50,** 459–463.

9. Randerath, L. & Randerath, E. (1967) in *Methods in Enzymology,* eds. Colowick, S. P. & Kaplan, N. O. (Adademic Press, New York), Vol. XII A, pp. 323–347.

10. Pettijohn, D. & Hanawalt, P. (1964) *J. Mol. Biol.* **9,** 395–410.

Relation between Nuclear and Chloroplast Genomes

Nuclear Dependency of Chloroplast Proteins in *Acetabularia*

Klaus APEL and Hans-Georg SCHWEIGER

It is now well established that chloroplasts contain all components necessary for genetic autonomy [1]. However, it is not fully understood to what degree this autonomy is actually accomplished. That the organelles are integrated in the cell has been shown by genetic experiments proving that the formation of many plastid properties is directed by the nucleus [2].

By estimating the amount of DNA per chloroplast of *Acetabularia*, it could be demonstrated that only a limited number of chloroplast proteins may be coded by this DNA. Therefore, most of the chloroplast proteins must get their genetic information from the nuclear DNA [3]. Direct evidence in favour of this idea has been obtained from experiments showing the transformation of the isozyme pattern of two enzymes bound to the chloroplasts after heterologous nuclear transplantation [4,5].

Thus it has been proved that at least two of the chloroplast proteins are determined by the nuclear DNA. The question arises, whether there are proteins which are coded by the chloroplast DNA. If this is the case, one should expect to find them among the structural proteins, since these are organelle-specific components.

Unusual Abbreviation. Triton X-100, octylphenoxypoly-ethoxyethanol.

This study was undertaken to investigate the origin of genetic information for chloroplast structural proteins. The insoluble chloroplast-proteins of two species of *Acetabularia* were examined for differences and since such differences could be demonstrated, it was possible to test by heterologous nuclear transplantation and implantation techniques whether the species-specific insoluble chloroplast-proteins are coded by the nuclear DNA or by the chloroplast DNA.

Furthermore, the two antibiotics cycloheximide and chloramphenicol were used to approximate the site of synthesis of the insoluble chloroplast-proteins. Cycloheximide has been shown to inhibit protein synthesis by 80 S ribosomes of eucaryotes [6], whereas the synthesis of proteins by 70 S ribosomes of bacteria [7] and chloroplasts [8] is insensitive to this antibiotic. In contrast, chloramphenicol inhibits protein synthesis by bacteria and isolated chloroplasts [9] but not that by 80 S ribosomes.

MATERIAL AND METHODS

Cells

Acetabularia mediterranea and *Acetabularia calyculus* were grown under standard conditions [3,10].

Transplantation and Implantation Techniques

Transplants were produced by grafting the rhizoid, containing the nucleus of one cell, to the stalk or the basal part of a stalk of another cell [3].

Besides the transplants, implants were used in some experiments. For this purpose the cell sap together with the nucleus was squeezed out of rhizoids by means of tweezers in an ice-cold sucrose solution (8 %) on a micro culture slide. Under a binocular microscope the rather large nucleus (0.1 mm) can be distinguished from polyphosphate granules. The nucleus was separated from adhering plasma material and injected into an anucleate basal part.

For the experiments only those implants were used in which regeneration of the cell was induced by the nucleus.

Isolation and Purification of Chloroplasts

Chloroplasts were prepared from 100 cells by means of the centrifugation technique [11]. The cells were centrifuged for 10 min at $650 \times g$. The cell plasma was collected in 5 ml chloroplast-buffer pH 6.1 [12]. The supernatant was discarded and the sediment was resuspended in 2 ml chloroplast-buffer. The suspension, which consisted of unwashed chloroplasts, was centrifuged for 10 min at $1500 \times g$, the supernatant was discarded and the sediment was resuspended again in 2 ml chloroplast-buffer. This suspension of washed chloroplasts was used for the experiments *in vitro*. All steps were carried out at $+2 \,°C$.

In order to purify the chloroplasts a crude extract of 150 plants was subjected to linear sucrose-density-gradient centrifugation ($50\,\%-25\,\%$ sucrose; w/v). The sucrose was dissolved in chloroplast-buffer according to Jensen and Bassham [12]. The cell plasma was suspended in 1 ml chloroplast-buffer ($+2\,°C$) and layered on the gradient. It was centrifuged for 70 min at $58000 \times g$ (SW 25 rotor) and $0\,°C$. After the run the gradient was fractionated and the chlorophyll content of each fraction was estimated by measuring the absorbance at 652 nm and the sucrose concentration was determined by the refractive index of each fraction (Fig. 1). The fractions of the chlorophyll peak (fraction 8—17) were pooled and the purified chloroplasts were isolated by recentrifugation.

Isolation of Chloroplast Proteins

Chloroplasts isolated from 10 cells yielded sufficient amounts of material for one experiment. The cell plasma was obtained by the centrifugation technique [11] and was collected in 1 ml 0.2 M citrate-phosphate buffer pH 3.7. The clear supernatant was discarded. The green pellet was resuspended in 1 ml 0.5 mM citrate-phosphate buffer pH 3.7 by agitation for 20 sec, centrifuged for 15 sec at $1000 \times g$ and the light-green supernatant discarded again. The sediment was suspended in 1.5 ml 0.5 mM citrate-phosphate buffer pH 3.7, $0.5\,\%$ Triton X-100. During the following 30 min the suspension was shaken twice for several seconds and then centrifuged for 10 min at $1300 \times g$. The dark-green supernatant was discarded.

The "insoluble" protein fraction of the sediment was dissolved in 0.025 ml of a mixture of phenol—formic acid—water (2:1:1, by vol.) [13] and after

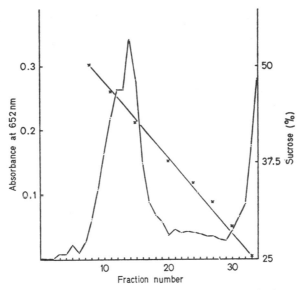

Fig. 1. *Sucrose-density-gradient centrifugation of the cell plasma of 150* A. mediterranea *cells for 70 min, at 0 °C and 58 000 × g (SW 25)*. In each fraction the sucrose concentration (×——×) and the chlorophyll content, measured by absorbance at 652 nm (●——●) were determined

150 min it was centrifuged for 10 min at $1300 \times g$. The proteins in the supernatant were called "chloroplast-proteins". All steps were carried out at 20 °C.

Electrophoretic Separation of the Proteins

Electrophoresis was performed according to a modification of the method of Biedermann and Drews [14]. The gel was prepared by addition of 4.075 g acrylamide, 0.25 ml ethylene diacrylate, 0.027 g ammonium persulfate and 0.25 ml diaminopropionitrile to 50 ml distilled water. During electrophoresis the gel (6 × 10 cm) was cooled from both sides. In order to get equal ion-distribution in the gel the voltage was applied for 2 h before the sample was run.

The electrophoretic separation was performed at 700 V (30 V/cm gel) and 11 mA and was stopped after 270 min. Two samples were separated simultaneously on one slab. After electrophoresis the gels were stained with an amido black solution (5 g per litre of 2 % acetic acid) for 1 h at room temperature and destained in 2 % acetic acid. The stained bands were scanned for absorbance at 578 nm.

After separation of the labelled proteins the gels were frozen on dry ice for 1 h. For analysis of the distribution of the radioactivity in the frozen gel, 1 mm slices were cut and measured for radioactivity.

Incorporation of Amino Acids

For the experiments with whole cells the plants were purified mechanically. Two h after purification the plants were transferred to Petri dishes which contained 20 ml Erd-Schreiber-solution [10]. The incorporation was started by the addition of the labelled amino acids to the medium. During the incubation the cells were illuminated with white light (2500 lux) at 20 °C. The incubation was stopped by washing the cells three times with Erd-Schreiber-solution which contained 0.1 % unlabelled amino acid. The chloroplast-protein fraction was isolated as described above.

For the experiments with isolated chloroplasts either [14C]leucine (final concentration: 0.25 μCi/ml; specific activity: 6.6 mCi/mmol) or a 14C-labelled amino acid mixture (final concentration: 3 and 6 μCi/ml, respectively; specific activity: 52 mCi/milliatom) was added to 4 ml chloroplast-buffer. The incorporation was started by the addition of 2 ml chloroplast suspension corresponding to 100 cells. The incubation was performed for 2 h at 25 °C under white light in a Warburg apparatus. Samples of 1 ml each were taken from the suspension and were pipetted quickly into 8 ml ice-cold 0.2 M citrate-phosphate buffer pH 3.7. The chloroplasts were sedimented at $2000 \times g$ for 2 min and the pellets were treated further for the isolation of the protein fraction. In some experiments a mixture of 18 unlabelled amino acids necessary for protein synthesis (final concentration: 25 μmol/ml each, free of [12C]leucine) together with [14C]leucine was dissolved in 4 ml chloroplast-buffer and was added to the chloroplast suspension.

Estimation of Radioactivity

The radioactivity of the 14C-labelled proteins was determined either directly in the phenol—formic acid—water solution or after electrophoresis in 1 mm slices of the gel.

10 ml Bray solution [15] was added to each sample and the counting was performed in a Packard Liquid

Scintillation Spectrometer. Under standard conditions the efficiency was 70 % and 59 % in the phenolic solution and in the slices, respectively.

Assays

Chlorophyll determinations were performed according to Arnon [16]. Since the proteins dissolved in the phenol—formic acid—water mixture could not

Table 1. *Dissolving of the insoluble chloroplast proteins*

Sample		Nitrogen	
		Amount	Fraction of total
		μg/100 plants	%
I	Total nitrogen of the membrane fraction after treatment with Triton X-100	164	100
II	Dissolved amount of nitrogen in the supernatant after treatment with phenol-formic acid-water (chloroplast proteins)	125	77
III	Amount of nitrogen in the sediment after treatment with phenol-formic acid-water	38	

be estimated directly, the samples were digested with sulfuric acid and the nitrogen content was measured by the phenol-hypochlorite method [17].

RESULTS

Dissolving of the Chloroplast Proteins

After treating the isolated chloroplasts of *Acetabularia* with Triton X-100, a fraction was obtained which did not dissolve in water. As was shown by quantitative experiments 75 % of the material was dissolved in the phenol—formic acid—water mixture (Table 1).

Electrophoretic Separation of the Chloroplast Proteins

The chloroplast proteins were dissolved in the phenol—formic acid—water mixture and were subjected to gel electrophoresis. Protein patterns from different cultures of the same species did not differ essentially. Whereas normally the chloroplast-proteins were isolated from a crude chloroplast preparation, in some cases the chloroplasts were purified by means of the sucrose-density-gradient centrifugation. The

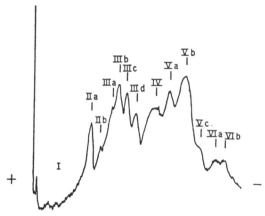

Fig. 2. *Electrophoresis of chloroplast proteins prepared from* A. mediterranea. The proteins isolated from 10 plants were subjected to electrophoresis as described in the Material and Methods section. After staining with amido black the gel was scanned at 578 nm. The peaks of the different sections (I—VI) were numbered sequentially according to their migration distance from the origin of the gel (+)

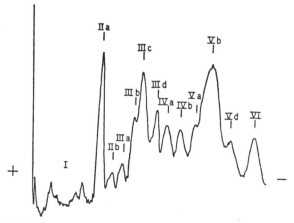

Fig. 3. *Electrophoretic analysis of chloroplast-proteins prepared from* A. calyculus. Conditions were as in Fig. 2

separation patterns in each case were identical. So it seems to be evident that the chloroplast proteins under investigation originated in chloroplasts and not in other cell components.

The Protein Pattern of Acetabularia mediterranea

The protein pattern of *A. mediterranea* displayed at least 12 peaks which could be classified in 5 sections (II—VI) (Fig. 2).

The densitometric tracing of section I displayed small peaks which did not reflect separated protein bands, but rather were due to the fact that during the application small amounts of the sample spread on the surface of the gel. Section II was subdivided into a more prominent peak IIa and a small one IIb. Whereas section III was composed of at least the four peaks IIIa—IIId, in section IV no further resolution could be achieved. Peaks Va and Vb were separated clearly from each other. Peak Vc appeared as a shoulder which could not be separated from the prominent peak Vb even by prolonging the time of electrophoresis. The resolution of the fastest running peaks VIa and VIb, which were separated from the section V, becomes obvious in Fig. 2.

The Protein Pattern of A. calyculus *and Comparison with that of* A. mediterranea

The electrophoretic pattern of the chloroplast-proteins of *A. calyculus* resembled those isolated from *A. mediterranea* (Fig. 3). The peaks IIa and IIb, IIIa, b, c and d and Vb were found in both species. Differences between the protein patterns of these two species were detected in sections IV, V and VI.

In contrast to *A. mediterranea* whose section IV appeared as a single peak, section IV of *A. calyculus* is characterized by the distinct resolution into the components IVa and b. If a mixture of chloroplast-proteins of *A. mediterranea* and *A. calyculus* was subjected to electrophoresis the two *A. calyculus* peaks disappeared in an unresolved broad peak IV. Probably this broad peak was due to the fact that the resolving capacity of the electrophoresis method was not sufficient to separate the components IV, IVa and b. As opposed to the *A. mediterranea* pattern, the *A. calyculus* pattern is characterized by the Va peak which is localized closely to peak Vb. However, the two bands Va of *A. mediterranea* and *A. calyculus* coincide in a mixed run.

An important difference between the two patterns existed in the fact that in the *A. calyculus* pattern, instead of the small peak Vc which was absent, a new peak Vd appeared which could be distinguished from the *A. mediterranea* peak Vc in a mixed run.

A further substantial difference was that instead of the paired peaks VIa and VIb in *A. mediterranea* only a single peak was found in *A. calyculus*. In a mixed run this peak VI fused with peaks VIa and b to give a single broad peak VI.

Transplants

Transplants were produced by grafting a rhizoid of *A. mediterranea*, which contained the nucleus, onto the anucleate basal part of a cell of *A. calyculus* ($med_1 cal_0$ transplant) and *vice versa* by transplanting a rhizoid of *A. calyculus* onto the anucleate basal part of an *A. mediterranea* cell ($cal_1 med_0$ transplant).

Six weeks after transplantation the chloroplast proteins were isolated and subjected to gel electrophoresis. It was found that the protein pattern of the $cal_1 med_0$ transplant had become similar to that of *A. calyculus* (Fig. 4). The prominent section IV of *A. mediterranea* had been resolved into the two peaks IVa and b which are specific for *A. calyculus*. The peaks Vc, VIa and b had disappeared. They had been replaced by the characteristic *A. calyculus* peaks Vd and VI which were distinctly separated from the peak Vb. This result suggests that under the influence of the *A. calyculus* rhizoid the chloroplast proteins of the anucleate *A. mediterranea* part became characteristic of *A. calyculus*.

In the parallel experiment with $med_1 cal_0$ transplants we studied the influence of the $med_1 cal_0$ on the chloroplast proteins of *A. calyculus* (Fig. 5). These experiments again showed that six weeks after transplantation, the original *A. calyculus* pattern had changed into an *A. mediterranea* pattern. The broad peak IV, the peak Vc, and the paired peaks VIa and b, were species-specific characteristics of *A. mediterranea*.

These results suggest that under the influence of the rhizoid the electrophoretic pattern of the chloro-

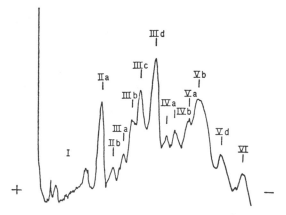

Fig. 4. *Electrophoretic analysis of chloroplast proteins prepared from* cal_1 med_0 *transplants.* Conditions were as in Fig. 2

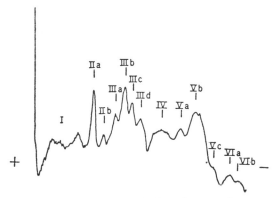

Fig. 5. *Electrophoretic analysis of chloroplast proteins prepared from* med_1 cal_0 *transplants.* Conditions were as in Fig. 2

plast proteins is changed. On the one hand, peaks which are specific for the species of the cytoplasm disappear, and on the other hand, peaks which are specific for the species of the rhizoid become manifest. Both effects indicate that it is the cell nucleus which governs the patterns of the chloroplast proteins.

Implants

In order to exclude the possibility that cellular components other than the nucleus, which were localized in the rhizoid together with the nucleus, affected the transformation of the protein pattern, cell nuclei were isolated and implanted into anucleate basal-cell fragments. Isolated nuclei of *A. mediterranea* were transferred into anucleate basal parts of *A. calyculus* ($med_1 \ cal_0$ implant) and *vice versa*, isolated nuclei of *A. calyculus* into anucleate basal parts of *A. mediterranea* ($cal_1 \ med_0$ implant). After six weeks in both cases the protein patterns of the implants resembled those of the corresponding transplants.

Table 2. *Influence of ^{14}C-labelled bacteria on the activity of the insoluble chloroplast-protein fraction of* A. mediterranea Incubation was performed with [14C]leucine under standard conditions (20 ml Erd-Schreiber-solution, 0.5 μCi/ml, [14C]-leucine, specific activity 6.6 mCi/mmol). Radioactivity is expressed per μg chlorophyll

Sample	Radioactivity
	counts $\times \text{min}^{-1} \times \mu\text{g}^{-1}$
I Insoluble chloroplast proteins of 7 cells which were incubated with [14C]leucine for 3 h	592
II Insoluble chloroplast proteins of 7 unlabelled cells which were centrifuged together with 7 labelled cells (incubated with [14C]leucine for 3 h). The rhizoids of only the unlabelled plants had been cut off	1.3
III Insoluble chloroplast proteins of 7 unlabelled plants	0.7

Incorporation of Labelled Amino Acids into the Chloroplast Proteins

Cellular protein synthesis was investigated by the incorporation of [14C]labelled amino acids into the protein. First of all it was necessary to exclude the possibility that incorporation was due to microbial contamination. In non-sterile cultures of *Acetabularia* cells such contamination may be attached to the outer surface of the cell wall and it is difficult to remove completely by mechanical purification. One way to avoid such difficulties is to extrude the cytoplasm out of the cell by centrifugation after removal of the rhizoid [11]. Under these conditions the cell wall is not destroyed.

The efficiency of this method was tested in the

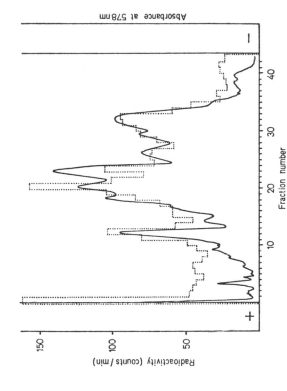

Fig. 6. *Absorbance of amido-black-stained insoluble chloroplast proteins of A. mediterranea after gel electrophoresis* (———) *and the distribution of the incorporated radioactivity in vivo along the gel* (------). Ten plants were incubated with 2.5 μCi [14C]-glycine (specific activity: 15.1 mCi/mmol; 0.125 μCi/ml Erd-Schreiber-solution) for 20 h

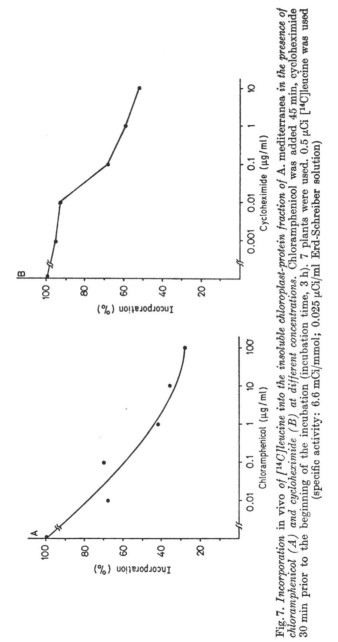

Fig. 7. *Incorporation* in vivo *of [¹⁴C]leucine into the insoluble chloroplast-protein fraction of A. mediterranea in the presence of chloramphenicol (A) and cycloheximide (B) at different concentrations.* Chloramphenicol was added 45 min, cycloheximide 30 min prior to the beginning of the incubation (incubation time, 3 h). 7 plants were used. 0.5 µCi [¹⁴C]leucine was used (specific activity: 6.6 mCi/mmol; 0.025 µCi/ml Erd-Schreiber solution)

following experiments (Table 2):

Seven cells incubated with labelled amino acids (sample I), seven cells incubated with unlabelled amino acids (sample III), and a combination of seven labelled together with seven unlabelled cells (sample II) were centrifuged. In sample II only the rhizoids of the unlabelled cells had been cut off, so that the cell plasma of the unlabelled plants was collected at the bottom of the centrifuge tube together with that material which had been removed by centrifugation from the surface of both the unlabelled and the labelled plants. The chloroplast protein fraction was isolated from the three samples and the radioactivity was estimated. It can be concluded that possible microbial contamination attached to the cell surfaces would not contaminate the chloroplast protein fraction.

Whole Cells

During the incubation of whole cells [^{14}C]leucine was incorporated at a constant rate for 3 h. Then the rate decreased. In our experiments the incubation time was 20 h. The electrophoretic pattern of the radioactive proteins isolated from labelled cells was found to be reproducible. At least three peaks could be separated distinctly (Fig. 6) in the fractions 10 to 15 (peak 1), 18—23 (peak 2) and 29—35 (peak 3). There was good agreement between the curves obtained by staining with amido black and by testing for radioactivity (Fig. 6).

The incorporation of [^{14}C]leucine into the insoluble chloroplast-protein fraction was inhibited by cycloheximide as well as by chloramphenicol. This result indicates that the chloroplast proteins were not synthesized exclusively either by 80 S or by 70 S ribosomes (Fig. 7).

In other experiments the effect of cycloheximide and of chloramphenicol on the incorporation of radioactive amino acids into the electrophoretically separated bands of the chloroplast proteins was studied.

For this purpose cycloheximide together with [^{14}C]glycine was added to the incubation medium up to a final concentration of 0.05 µg/ml. The labelled proteins were separated electrophoretically and the incorporation pattern of the plants treated with cycloheximide was compared with that of untreated control plants (Fig. 8). In the presence of cycloheximide the incorporation into peak 3 was inhibited

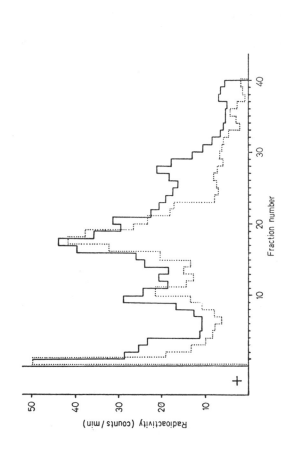

Fig. 8. *Electrophoretic pattern of the radioactivity of the insoluble chloroplast-protein fraction of A. mediterranea incorporated in vivo.* Ten plants at a time were incubated in the presence (·····) or in the absence (——) of cycloheximide (0.05 µg/ml) with 1 µCi [^{14}C]glycine (specific activity: 15.1 mCi/mmol; 0.05 µCi/ml Erd-Schreiber solution) for 20 h

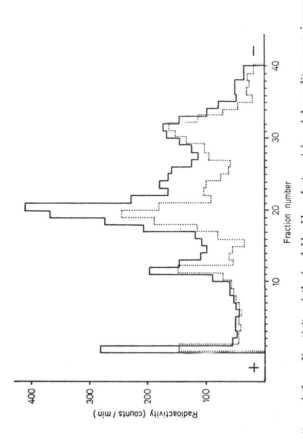

Fig. 9. *Electrophoretic pattern of the radioactivity of the insoluble chloroplast proteins of* A. mediterranea *incorporated in vivo.* Ten plants at a time were incubated in the presence (······) or in the absence (———) of chloramphenicol (10 μg/ml) with 5 μCi [^{14}C]glycine (specific activity: 15.1 mCi/mmol; 0.25 μCi/ml Erd-Schreiber solution) for 20 h

almost completely, whereas peaks 1 and 2 showed a distinct labelling. However, if compared with the control cells, this incorporation was inhibited too, namely by 51 % and 10 % in peaks 1 and 2, respectively.

Chloramphenicol was added to the incubation medium together with [^{14}C]glycine up to a concentration of 10 μg/ml. The incorporation patterns of treated and untreated cells were compared (Fig. 9). While there was almost no effect on the labelling of the cycloheximide-sensitive peak 3 (15.5 % inhibition), there was an obvious inhibition of the incorporation into peaks 1 and 2, namely by 59 % and 50 %, respectively.

Isolated Chloroplasts

Isolated chloroplasts were capable of incorporating [^{14}C]leucine into the insoluble protein fraction only if an amino acid mixture was added (Table 3). An in-

Table 3. *The incorporation of ^{14}C-labelled amino acids into the insoluble protein fraction of isolated chloroplasts*
The samples were incubated for 2 h under standard conditions as described in the Material and Methods section. The total volume of the incubation medium was 6 ml. Radioactivity is expressed per μg chlorophyll

Additions to incubation medium	Radioactivity
	counts $\times min^{-1} \times \mu g^{-1}$
[^{14}C]Leucine (0.25 μCi/ml, specific activity: 6.6 mCi/mmol)	0.12
[^{14}C]Leucine + unlabelled-amino acids, free of [^{12}C]leucine (final concentration of each amino acid 25 μmol/ml)	16
[^{14}C]Leucine + unlabelled amino acid mixture + ATP and GTP (1 μmol/ml)	15.5
^{14}C-labelled amino acid mixture (3 μCi/ml, specific activity 52 mCi/milliatom)	592

cubation time of 2 h was insufficient to get a definite distribution pattern of radioactivity after the gel electrophoresis. Addition of ATP and GTP did not increase the incorporation rate. When a ^{14}C-labelled amino acid mixture was used, the radioactivity

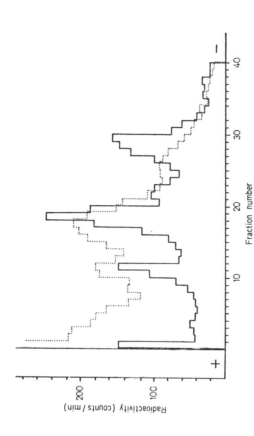

Fig. 10. *Electrophoretic pattern of the in vivo* (———) *and the in vitro* (······) *incorporated radioactivity of the insoluble chloroplast proteins of A. mediterranea.* Ten plants were incubated with 5 μCi [14C]glycine (specific activity 15.1 mCi/mmol; 0.25 μCi/ml Erd-Schreiber solution) for 20 h. Chloroplasts isolated from 100 plants were incubated with 6 μCi 14C-labelled amino acid mixture (specific activity 52 mCi/milliatom; 1 μCi/ml chloroplast buffer) for 2 h

incorporated by isolated chloroplasts was such that after the electrophoretic separation of the chloroplast proteins a distribution pattern could be revealed. The incorporation patterns of the chloroplast proteins obtained from both isolated chloroplasts and whole plants were compared (Fig. 10). Although the amido black-stained protein patterns obtained in each case compared well, the incorporation patterns differed distinctly with respect to peak 3. This peak was synthesized only in whole cells and was absent in the incorporation pattern of isolated chloroplasts. Incorporation into peaks 1 and 2 could be demonstrated in whole cells as well as in isolated chloroplasts.

The fact that incorporation into peak 3 was observed only in whole cells but not in isolated chloroplasts is presumably related to the removal of the 80 S ribosomes during the isolation of the chloroplasts. In this connection it is not surprising that the incorporation pattern *in vivo* in the presence of cycloheximide resembled the incorporation pattern of isolated chloroplasts, since this inhibitor exerts its effect on 80 S ribosomes.

DISCUSSION

The electrophoretic patterns of the chloroplast proteins of *A. mediterranea* and *A. calyculus* demonstrate the presence of a number of common proteins and of some proteins which are species-specific. Similar conclusions have been drawn by Biedermann and Drews [14]. They dissolved the thylakoid proteins of different species of purple bacteria and after electrophoretic separation they were able to demonstrate species-specific protein patterns for each of the investigated species. In a similar way Mani and Zalik [18] electrophoretically resolved protein fractions prepared from chloroplast membranes of *Phaseolus* and *Triticum* and they also found species-specificity of the protein patterns.

Contrary to these findings Braunitzer and Bauer [13] were not able to find differences between the chloroplast membrane proteins from a monocotyledon, a dicotyledon and a green alga. However, they could resolve their samples only into three components, whereas in the experiments reported here at least 12 bands were separated. A similar high number of protein bands was found by Hoober [19] after the electrophoretic separation of chloroplast membrane proteins of *Chlamydomonas*.

The species-specificity of some of the chloroplast proteins in *Acetabularia* made it possible to demonstrate the influence of the nucleus on the formation of these proteins by means of heterologous nuclear transplantation and implantation techniques. Six weeks after the operation the species-specific chloroplast proteins had been replaced by those which correspond to the donor species of the nucleus. These changes cannot be explained by variations in the composition of the medium or by other external factors. Furthermore, the changes cannot be due to the transfer of contamination by cellular components other than the nucleus as might be considered in the case of the transplantation experiments. By the implantation experiments this possibility is definitely ruled out. Therefore it is justified to state that it is the nucleus which effects the transformation of the pattern of the chloroplast proteins both in the transplantation experiments and in the implantation experiments.

Other experimental results have shown that in *Acetabularia* the number of chloroplasts is doubled within 15 days [20]. This means that within six weeks most of the chloroplasts including the chloroplast proteins are newly formed [21]. Based on these observations one may expect that within six weeks, in the transplants, most of the chloroplast membrane proteins have been synthesized unter the control of the heterologous nucleus. However, besides the grafted nucleus the mRNA remaining from the acceptor cell nucleus must be considered to be involved in the protein synthesis of the grafts.

The synthesis of nucleus-dependent proteins is continued in anucleate cells and in cell fragments. It is obviously directed by stable mRNA molecules. However, since in heterologous grafts the pattern of chloroplast proteins is changed, the conclusion must be drawn that under the influence of the implanted heterologous nucleus the already existing mRNA of the acceptor cell nucleus is no longer involved in translation and most probably is preferentially degraded. Similar conclusions have been drawn from experiments on two enzymes in *Acetabularia* [4,5].

Chloroplasts have been shown to contain all components involved in protein synthesis [3]. The incorporation of amino acids [22,23], RNA precursors [24] and DNA precursors [25] into the chloroplasts has been demonstrated. Thus chloroplasts possess all the qualifications for genetic autonomy. However, it has been found earlier that the autonomy is only partial. For instance the malic and the lactic dehydrogenase, both associated with chloroplasts, are coded by the cell nucleus [4,5]. It also has been shown, by estimating the amount of DNA per chloroplast, that the chloroplast proteins could be coded only partially by this DNA [3].

In some ways it is surprising that the insoluble as well as the soluble proteins of the chloroplasts are nucleus-dependent. The insoluble proteins play a major part in the architecture of the membrane system of the chloroplast. Since they are organelle-specific components of the chloroplasts, one would expect that they are synthesized under the control of the plastome. However, our results prove that the chloroplast is integrated into the cell at least to such a degree, that the nuclear DNA codes the formation of chloroplast-specific structural components as well as the synthesis of soluble enzyme proteins. This statement applies to those proteins of the membrane fraction which turned out to be species-specific. This result does not exclude the possibility that some or all of the remaining insoluble chloroplast proteins are coded by the chloroplast DNA.

The experiments with chloramphenicol and cycloheximide clearly demonstrate that a protein synthesizing system is effective within the chloroplast.

In accordance with the results of Hoober *et al.* [26] and Hoober [19] it may be concluded from our experiments that some of the insoluble chloroplast proteins are synthesized on 80 S and others on chloroplast ribosomes. The inhibition of incorporation into peaks 1 and 2 by both cycloheximide and chloramphenicol is difficult to interpret. It may be that these peaks contain several labelled unresolved protein fractions which are synthesized either by 80 S or by chloroplast ribosomes. The absence of peak 3 in the incorporation pattern of whole cells incubated in the presence of cycloheximide could also be observed in the incorporation experiments with isolated chloroplasts after the removal of 80 S ribosomes without the inhibitor. From these results it can be concluded that the labelling of peak 3 takes part on 80 S ribosomes.

The inhibiting effect of chloramphenicol on the

labelling of peaks 1 and 2 seems to indicate that some chloroplast proteins are synthesized on chloroplast ribosomes. However, there is no direct evidence that these proteins are coded by the chloroplast DNA. One should take into consideration the possibility that protein synthesis on chloroplast ribosomes may obtain information from the nuclear DNA.

REFERENCES

1. Gibor, A., and Granick, S., *Science (Washington)*, 145 (1964) 890.
2. Kirk, J. T. O., and Tilney-Bassett, R. A. E., *The Plastid: their Chemistry, Structure, Growth and Inheritance*, W. H. Freeman, London and San Francisco, 1967.
3. Schweiger, H. G., *Curr. Top. Microbiol. Immunol.* 50 (1969) 1.
4. Reuter, W., and Schweiger, H. G., *Protoplasma*, 68 (1969) 357.
5. Schweiger, H. G., Master, R. W. P., and Werz, G., *Nature (London)*, 216 (1967) 554.
6. Siegel, M. R., and Sisler, H. D., *Biochim. Biophys. Acta*, 87 (1964) 83.
7. Ennis, H. L., and Lubin, M., *Science (Washington)*, 146 (1964) 1474.
8. Bonotto, S., Goffeau, A., Janowski, M., Vanden Driessche, T., and Brachet, J., *Biochim. Biophys. Acta*, 174 (1969) 704.
9. Ellis, R. J., *Science (Washington)*, 163 (1969) 477.
10. Hämmerling, J., *Ann. Rev. Plant Physiol.* 14 (1963) 65.
11. Schweiger, H. G., *Planta*, 68 (1966) 247.
12. Jensen, R. G., and Bassham, J. A., *Proc. Nat. Acad. Sci. U. S. A.* 56 (1966) 1095.
13. Braunitzer, G., and Bauer, G., *Naturwissenschaften*, 54 (1967) 70.
14. Biedermann, M., and Drews, G., *Arch. Mikrobiol.* 61 (1968) 48.
15. Bray, G. A., *Anal. Biochem.* 1 (1960) 279.
16. Arnon, D. I., *Plant Physiol.* 24 (1949) 1.
17. Bremer, H. J., and Schweiger, H. G., *Planta*, 55 (1960) 13.
18. Mani, R. S., and Zalik, S., *Biochim. Biophys. Acta*, 200 (1970) 132.
19. Hoober, J. K., *J. Biol. Chem.* 245 (1970) 4327.
20. Clauss, H., Lüttke, A., Hellmann, F., and Reinert, J., *Protoplasma*, 69 (1970) 313.
21. Clauss, H., *Planta*, 52 (1958) 334.
22. Goffeau, A., and Brachet, J., *Biochim. Biophys. Acta*, 95 (1965) 302.
23. Goffeau, A., *Biochim. Biophys. Acta*, 174 (1969) 340.
24. Berger, S., *Protoplasma*, 64 (1967) 13.

25. Berger, S., and Schweiger, H. G., *Physiol. Chem. Phys.*
    1 (1969) 280.
26. Hoober, J. K., Siekevitz, P., and Palade, G. E., *J. Biol.
    Chem.* 244 (1969) 2621.

A MAJOR POLYPEPTIDE OF CHLOROPLAST

MEMBRANES OF *CHLAMYDOMONAS REINHARDI*

Evidence for Synthesis in the

Cytoplasm as a Soluble Component

J. KENNETH HOOBER

ABSTRACT

Electrophoresis of thylakoid membrane polypeptides from *Chlamydomonas reinhardi* revealed two major polypeptide fractions. But electrophoresis of the total protein of green cells showed that these membrane polypeptides were not major components of the cell. However, a polypeptide fraction whose characteristics are those of fraction *c* (a designation used for reference in this paper), one of the two major polypeptides of thylakoid membranes, was resolved in the electrophoretic pattern of total protein of green cells. This polypeptide could not be detected in dark-grown, etiolated cells. Synthesis of the polypeptide occurred during greening of etiolated cells exposed to light. When chloramphenicol (final concentration, 200 μg/ml) was added to the medium during greening to inhibit chloroplastic protein synthesis, synthesis of chlorophyll and formation of thylakoid membranes were also inhibited to an extent resulting in levels of chlorophyll and membranes 20–25% of those found in control cells. However, synthesis of fraction *c* was not affected by the drug. This polypeptide appeared in the soluble fraction of the cell under these conditions, indicating that this protein was synthesized in the cytoplasm as a soluble component. When normally greening cells were transferred from light to dark, synthesis of the major membrane polypeptides decreased. Also, it was found that synthesis of both subunits of ribulose 1,5-diphosphate carboxylase was inhibited by chloramphenicol, and that synthesis of this enzyme stopped when cells were transferred from light to dark.

INTRODUCTION

Cells of the *y-1* strain of *Chlamydomonas reinhardi*, a unicellular green alga, are capable of growth for several days in the absence of light, but under these conditions fail to synthesize chlorophyll and assemble thylakoid membranes within the chloroplast (1, 2). The membranes are diluted among the daughter cells, and after growth for 4–5 days in the dark in liquid culture, the cellular contents of chlorophyll and thylakoid membranes are less than 5% of those in cells grown in continuous light (2). When etiolated cells are returned to light, chlorophyll, thylakoid membranes, and the functional activities of these membranes increase in parallel (3, 4).

Thylakoid membranes are assembled within the chloroplast from proteins synthesized both in the

67

cytoplasm and in the chloroplast (5, 6). Two major polypeptide fractions, as revealed by gel electrophoresis, are among those synthesized in the cytoplasm (5). These polypeptides must migrate into the chloroplast for membrane formation to occur. The mechanism by which this intracellular transport occurs is not known, but it is reasonable to assume that the membrane polypeptides are transported as soluble proteins. If synthesized in a soluble form, these polypeptides should appear in the soluble fraction of the cell when normal membrane assembly is prevented.

As described in this paper, inhibiting membrane formation by treating cells with chloramphenicol resulted in the accumulation of one of the major membrane polypeptides in the soluble fraction of the cell. The results also indicate that the enzyme ribulose 1,5-diphosphate carboxylase is synthesized in the chloroplast, and is the main product of protein synthesis in the organelle. Some proteins, including ribulose 1,5-diphosphate carboxylase, are synthesized very little or not at all when cells are transferred from light to dark.

METHODS

Handling of the Cells

Cells of *Chlamydomonas reinhardi y-1* were grown in light or dark as described before (2, 4). Etiolated cells, which contained 1 μg or less of chlorophyll per 10^7 cells after growth for 4 days in the dark, were suspended in fresh medium, supplemented with KH_2PO_4 (4), to a density of 6 $\times$ 10^6 cells/ml. 40-ml portions of the cell suspension were added to 500-ml Erlenmeyer flasks and exposed to light from white fluorescent lamps at an intensity of about 8000 lux at 25°C while on a rotating platform. At 0 hr, 4.4 ml of a solution of chloramphenicol (2 mg/ml) were added to treated cells, while an equal volume of distilled water was added to the control cells.

Preparation of Thylakoid Membranes

Thylakoid membranes were isolated as described previously (5), except that in some experiments the broken cell preparation (in 0.3 M sucrose, 25 mM Tris[1]-HCl, pH 7.6, and 1 mM MgCl$_2$) was incubated for 15 min at 4°C after adding pancreatic deoxyribonuclease to a final concentration of 10–20 μg/ml. This procedure was found to increase the yield of membranes and facilitate suspension of the crude

---

[1] *Abbreviations used:* Tris, tris(hydroxymethyl) aminomethane; EDTA, ethylenediaminetetraacetate; DCI, 2,6-dichloroindophenol; CAP, chloramphenicol.

membrane pellet after the first centrifugation step (5). In some experiments, the isolated membranes were suspended in either distilled water, 1.0 M Tris-HCl (pH 7.6), or 1 mM EDTA[1] (pH 8.0) and allowed to stand 30 min at 4°C. While in the wash solution, the membranes were further fragmented by passing the suspension through a French pressure cell at 5000 psi. The membranes were recovered by centrifuging the sample at 100,000 g for 40 min.

Cell Fractionation

In order to prepare samples of total cellular protein, total membrane protein, and soluble protein, cells were washed three times at 4°C with 0.3 M sucrose containing 10 mM Tris-HCl (pH 7.6) and broken by a passage through a chilled French pressure cell at 6000 psi. To a portion of the broken-cell preparation was added trichloroacetic acid to a final concentration of 10%. The ensuing precipitate provided a sample of total *cellular* protein. The remainder of the broken-cell preparation was centrifuged at 50,000 g_{av} for 15 min at 2°C. The pellet obtained provided a sample of total *membrane* protein. The 50,000 g supernatant fluid, which was free of chlorophyll-containing membranes, was then centrifuged for 3 hr at 120,000 g_{av} at 2°C. The supernatant fluid was removed, and trichloroacetic acid was added to a final concentration of 10%. The ensuing precipitate provided a sample of *soluble* protein. The precipitates, each containing 3–6 mg of protein, were washed with 5 ml of 5% trichloroacetic acid and finally with 2 ml of water. The samples were stored at -15°C if not used immediately.

Gel Electrophoresis

In preparation for electrophoresis, the sample of soluble protein was dispersed in water, and the suspension was made alkaline (pH 8–9) by adding a few drops of 2% Na_2CO_3. The volume of the sample was then measured, and an equal volume of a solution containing 0.2 M Tris acetate (pH 9.0), 0.7 mM EDTA, 1.0 M urea, and 4.0% sodium dodecyl sulfate was added. The final protein concentration was about 8 mg/ml.

Samples of total cellular protein were extracted with 90% acetone at room temperature so as to remove lipids. The insoluble protein was recovered by centrifugation at 1000 g for 5 min and then dissolved in a solution containing 0.1 M Tris acetate (pH 9.0), 0.35 mM EDTA, 0.5 M urea, and 2% sodium dodecyl sulfate to a final protein concentration of about 8 mg/ml.

Samples of membranes were extracted with 90% acetone at room temperature. After sedimenting the insoluble protein at 1000 g for 5 min, the acetone extract was removed and several grains of NaCl were added to the extract to bring any protein remaining in the extract out of solution. The extract was then added to the extracted residue, and the sample was again centrifuged. This step was found necessary for the recovery of all membrane protein, particularly from membranes washed with EDTA. The final, combined precipitates of membrane protein were dissolved in the same solution as given above for total cellular protein except that the concentration of sodium dodecyl sulfate was 1% and the final protein concentration was about 5 mg/ml. Before electrophoresis, each sample was treated with 2-mercaptoethanol (5). The procedures for gel electrophoresis, densitometry, and the determination of radioactivity in the gels were described previously (5). During the experiments and preparations for electrophoresis, corresponding samples from control and chloramphenicol-treated cells were handled identically. Thus, the results are equivalent on a cell basis.

Preparation of Fraction I Protein

Green cells were washed two times with 25 mM Tris-HCl (pH 7.6) containing 25 mM MgCl₂ and broken with a French pressure cell at 6000 psi. The sample was centrifuged for 15 min at 20,000 g and portions of the supernatant fluid were layered over linear gradients, 32 ml in volume, of 0.15–0.60 M sucrose containing 25 mM Tris-HCl (pH 7.6) and 25 mM MgCl₂. The gradients were centrifuged for 6 hr at 26,000 rpm at 2°C in a SW 27 rotor, and then analyzed with an ISCO (Instrumentation Specialties Co., Lincoln, Neb.) gradient fractionator. The region of the gradients containing the Fraction I protein was collected, and the protein was precipitated by adding trichloroacetic acid to a final concentration of 10%. The sample was then treated as described above for the sample of soluble protein in preparation for electrophoresis.

Assays

Chlorophyll was extracted with 80% acetone and measured spectrophotometrically (7, 8). Photoreduction of 2,6-dichloroindophenol was determined as described previously (4). Protein was estimated by the method of Lowry et al. (9) after the dissolving of acetone-extracted samples in 0.1 N NaOH.

Materials

Acrylamide and N,N'-methylene-bis-acrylamide were obtained from Eastman Organic Chemicals, Rochester, N.Y.; the acrylamide was recrystallized from chloroform (10). Omnifluor and arginine-³H (7.3 Ci/mmole) were purchased from New England Nuclear Corp., Boston, Mass. Triton-X100 was obtained from Beckman Instruments Inc., Fullerton, Calif. Chloramphenicol was provided by Parke-Davis. Deoxyribonuclease (Type I, ribonuclease-free) was obtained from Worthington Biochemical Corp., Freehold, N.J. Urea solutions were prepared fresh and passed through a mixed bed resin (Amberlite MB-1) before use. Concentrated stock solutions of urea, sucrose, and Tris (Sigma Chemical Co., St. Louis, Mo.) buffers were passed through a Millipore filter (Millipore Corp., Bedford, Mass.) after preparation. All other chemicals were of reagent grade.

RESULTS

Membrane Polypeptides

When etiolated cells of *Chlamydomonas reinhardi* y-1 were suspended in fresh growth medium and exposed to light, chlorophyll and thylakoid membranes within the chloroplast increased rapidly after a slow phase of about 3 hr. In work described previously (5), these membranes were isolated and the polypeptide components were

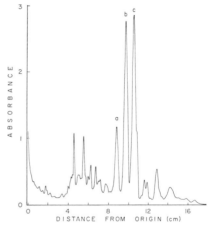

FIGURE 1 Electrophoretic analysis of the polypeptides of thylakoid membranes prepared from *C. reinhardi*. Membranes were isolated as before (5), washed with 1.0 M Tris-HCl, pH 7.6, extracted with 90% acetone, and prepared for electrophoresis as described under Methods. About 50 μg of the polypeptides (in a volume of 10 μl) were applied to the gel column and subjected to electrophoresis in the presence of sodium dodecyl sulfate (5) at 1.8 v/cm for 30 min and 6 v/cm for 7 hr. The gel was stained with Coomassie blue (5) and scanned at 563 mμ. The letters *a*, *b*, and *c* indicate fractions referred to in this paper.

examined by gel electrophoresis. Two major polypeptide fractions were resolved. Subsequently, the isolation procedure was extended to include a wash with 1.0 M Tris-HCl, pH 7.6, or with 1 mM EDTA, pH 8.0, conditions which reportedly extract some loosely bound proteins (11, 12). Fig. 1 shows the pattern of polypeptides from membranes washed with 1.0 M Tris-HCl. When compared to the pattern for membranes not washed after isolation (e.g., Fig. 2 B), the pattern in the washed sample showed only slight decreases in the amounts of some of the minor fractions. A pattern nearly identical to that shown in Fig. 1 was obtained after membranes were washed with 1 mM EDTA. The isolation procedure (5) apparently was sufficient for preparing membranes nearly free of extraneous proteins. In Fig. 1, the two prominent fractions marked b and c have molecular weights of 24,000 and 21,000, respectively (5), and together account for about 40% of the total protein stain on the gel. The assignments of the letters a, b, and c to the three fractions shown in Fig. 1 are for reference in this paper.

Identification of Membrane Polypeptides in Total Cellular Protein

Although electron microscopy revealed a large amount of thylakoid membrane material in green cells (2, 13), electrophoresis of total cellular protein did not reveal the membrane polypeptides as major protein components of these cells. Yet for subsequent experiments, it was necessary to establish that these polypeptides could be detected after electrophoresis of total protein. In Fig. 2, the pattern for total cellular protein (Fig. 2 A) is compared to that for the protein of isolated thylakoid membranes (Fig. 2 B). Several fractions (arrows) were resolved in the total protein sample which corresponded to the major polypeptides of the membranes. Whereas fractions containing a and c (see Fig. 1) appeared as discrete peaks in the total protein pattern, b appeared as a shoulder on the trailing side of another, larger fraction. The alignment of the patterns was confirmed by electrophoresis of a mixture of the membrane protein and the total protein.

So as to determine definitively whether the fractions in the total protein sample indicated by arrows in Fig. 2 were membrane components, the electrophoretic pattern for total protein of green cells was compared to that for yellow cells. Since yellow cells contain little thylakoid mem-

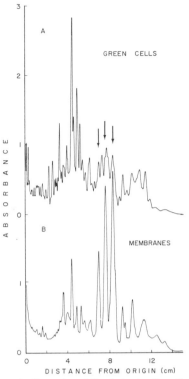

FIGURE 2 Comparison of the electrophoretic patterns for total polypeptides of green cells (Fig. 2 A) and for thylakoid membrane polypeptides (Fig. 2 B). Total protein was obtained, as described under Methods, from *C. reinhardi* cells grown 2 days in continuous light. Thylakoid membranes were isolated (5) from green cells. The two samples were extracted with 90% acetone and were prepared for electrophoresis as described under Methods. Portions of the samples containing about 50 μg of the polypeptides were subjected to electrophoresis on companion gel columns at 1.8 v/cm for 30 min and 7.2 v/cm for 4 hr. The arrows show the position, in the total polypeptide pattern, of the membrane fractions marked a, b, and c in Fig. 1. The alignment of the patterns was checked by scanning a gel containing a mixture of the two samples.

branes (2), a specific difference in the amounts of the membrane polypeptides might be revealed by such a comparison. As shown in Fig. 3, the pattern of protein from green cells (solid line) showed a peak corresponding to membrane fraction c (at 10 cm from the origin, also indicated

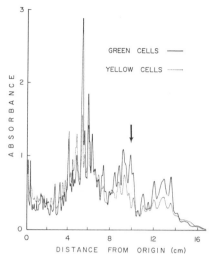

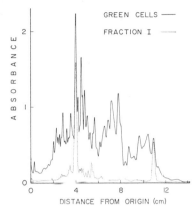

FIGURE 3 Comparision of the electrophoretic patterns for total polypeptides of green and yellow cells. Total protein from about 6×10^7 cells was obtained for each sample and treated in preparation for electrophoresis as described under Methods. Polypeptides in a volume of 10 μl were subjected to electrophoresis on each gel column at 1.8 v/cm for 30 min and 6 v/cm for 6.75 hr. The total area under the trace for green cells was about 35% greater than that for yellow cells, which is approximately the difference in protein content per cell between green and yellow cells (4, 6).

by arrow), but no corresponding peak was observed in the pattern of protein from yellow cells (dotted line). The two small peaks in this region of the pattern for yellow cells were present in the pattern for green cells as shoulders on either side of the peak for fraction *c*. A fraction containing membrane fraction *a* (at 8.5 cm from the origin) was also present in the pattern for green cells but not in that for yellow cells. Membrane fraction *b* was again less well resolved.

Identification of the Subunits of Ribulose 1,5-Diphosphate Carboxylase in the Total Protein Sample

With the exception of the region from 8 to 10.5 cm from the origin, the patterns shown in Fig. 3 were *qualitatively* similar although some fractions were present in different amounts in the two types of cells. The main polypeptide in both green and yellow cells migrated at a rate corresponding to

FIGURE 4 Comparision of the electrophoretic patterns for total polypeptides of green cells and of Fraction I protein. Total protein of green cells and Fraction I protein were obtained as described under Methods. The samples were treated as described for total protein, except that Fraction I protein was not extracted with acetone. Polypeptides in a volume of 10 μl were subjected to electrophoresis on each gel column at 1.8 v/cm for 30 min and 7.2 v/cm for 4.5 hr. The alignment of the patterns was checked by scanning a gel containing a mixture of the two samples.

a molecular weight of about 55,000 (the peak at 5.5 cm from the origin in Fig. 3). This is the size of the large subunit of ribulose 1,5-diphosphate carboxylase (14, 15). Also, the fraction at 13.5 cm contained polypeptides the size of the small subunit of this enzyme (14). This identification was supported by comparing the electrophoretic pattern for total protein of green cells with that of Fraction I protein, which is primarily ribulose 1,5-diphosphate carboxylase (16). As shown in Fig. 4, the prominent polypeptide in the sample from green cells migrated at the same rate as the large subunit of this enzyme (to 4 cm from the origin in this experiment). The identification of the small subunit (at 11 cm) in the total protein pattern was also confirmed. The results shown in Fig. 3, in agreement with measurements of enzymatic activity (2), indicated that the level of this enzyme in green cells was about two times that in the yellow cells.

Effects of Chloramphenicol on Formation of Membrane and Synthesis of Polypeptides

Since it was possible to resolve a major polypeptide of thylakoid membranes and the two subunits of ribulose 1,5-diphosphate carboxylase by gel electrophoresis of total cellular protein, experiments were done in order to find out how these proteins are affected when chloroplastic protein synthesis is inhibited. As was found earlier (5), the two major polypeptide fractions of the thylakoid membranes are synthesized in the cytoplasm and therefore synthesis of these proteins should not be affected when protein synthesis is inhibited in the chloroplast. However, if membrane assembly were concomitantly inhibited, these proteins might accumulate in a nonmembrane fraction.

Chloramphenicol is an inhibitor of protein synthesis on chloroplastic ribosomes (17–20) and, as a result, of chloroplast development (4, 21, 22). When etiolated *Chlamydomonas* cells were exposed to light in the presence of 200 μg of chloramphenicol/ml, the increase in chlorophyll was strongly inhibited compared to control cells (Fig. 5).

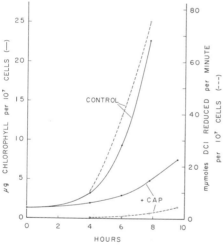

FIGURE 5 Effects of chloramphenicol at 200 μg/ml on the increases in chlorophyll and Hill reaction activity during greening of etiolated *C. reinhardi*. Yellow cells were suspended in fresh medium to 4×10^6 cells/ml and exposed to 8000 lux from white fluorescent lamps at 25°C. Each 50 ml portion of the suspension (in 500-ml flasks) received 5.5 ml of either chloramphenicol (2 mg/ml) or water at 0 hr. At the times indicated, control and chloramphenicol-treated samples were assayed for the ability to photoreduce DCI as described previously (4). Chlorophyll was measured spectrophotometrically (7) in 80% acetone extracts (———), chlorophyll; (----), photoreduction of DCI.

Furthermore, photoreduction of 2,6-dichloroindophenol was barely detectable in chloramphenicol-treated cells. Previous data indicated that proteins necessary for this activity of the membrane were synthesized on chloroplastic ribosomes (1). Examining cells with the electron microscope revealed an amount of thylakoid membranes in chloramphenicol-treated cells approximately in proportion to the amount of chlorophyll present (J. K. Hoober and G. E. Palade, unpublished results).

The magnitude of the chloramphenicol effect shown in Fig. 5 was dependent upon the time of addition of the drug. Results similar to that in Fig. 5 were obtained when chloramphenicol was added during the 1st hr of light exposure. If the drug was added after 3–4 hr of exposure to light, when the cells were actively making chlorophyll, no inhibition was observed until about 3 hr later, even though a clear effect was seen on chloroplastic ribosomes in less than 15 min under similar conditions (18). These results suggest that chloramphenicol affects chlorophyll synthesis indirectly, possibly by inhibiting the synthesis of catalytic proteins needed for chlorophyll biosynthesis.[2]

Since chloramphenicol does not affect cytoplasmic protein synthesis (17, 18), it can be assumed that the major membrane polypeptides are still synthesized in the presence of this drug. This was tested by the following experiment. Chloramphenicol was added to the culture medium to a concentration of 200 μg/ml at the time yellow cells were exposed to light. Arginine-^{3}H was then added at 5 hr, i.e. after control cells had begun to actively make thylakoid membranes, and 90 min later control and chloramphenicol-treated cells were collected. Total cellular protein samples were subjected to electrophoresis, and the patterns of radioactivity were determined.

---

[2] The effects of different concentrations of chloramphenicol should be emphasized. As found previously (4), chloramphenicol at 20–25 μg/ml caused only a small inhibition of chlorophyll synthesis. It was estimated that chloroplastic protein synthesis was inhibited about 70% under these conditions, and thus sufficient synthesis of enzymes possibly occurred to allow nearly normal rates of chlorophyll synthesis. Chloramphenicol at 200 μg/ml produced a maximal level of inhibition of protein synthesis for this drug (4), and also produced the effects on greening shown in Fig. 5. However, it is also possible that at this concentration the drug inhibited chlorophyll synthesis by means other than via its effects on protein synthesis.

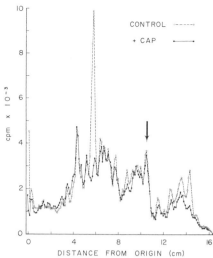

FIGURE 6 Electrophoretic analysis of the incorporation of arginine-^{3}H into polypeptides during greening
of *C. reinhardi* cells incubated with or without chloramphenicol (200 μg/ml). Yellow cells were suspended in
fresh medium to 6 × 10^6 cells/ml. Each 40 ml portion
(in 500 ml flasks) received 4.4 ml of either chloramphenicol (2 mg/ml) or water at 0 hr. After 5 hr in
light, arginine-^{3}H was added to each flask to a final
concentration of 1 μCi per ml, and 90 min later the cells
were collected, washed with 0.3 M sucrose containing
10 mM Tris-HCl, pH 7.6, and broken in the same buffer.
Total protein samples were prepared for electrophoresis as described under Methods, and portions of the
samples containing about 160 μg of the polypeptides in
a volume of 20 μl were applied to companion gel columns and subjected to electrophoresis at 2 v/cm for
20 min and 6 v/cm for 7 hr. After staining, the gels
were cut into 1-mm sections, each of which was incubated at 55°C overnight with 0.1 ml of 30% H$_2$O$_2$.
After cooling, 10 ml of a solution containing toluene,
Triton-X100, and Omnifluor (5) were added to each,
and the radioactivity was determined. The arrow
indicates the position in the gel of membrane fraction *c*.

Fig. 6 shows, firstly, that the labeling pattern for
control cells (dashed line) was related to the pattern of protein stain for green cells (Figs. 2 and
3). Thus, many proteins are synthesized during
the greening period, and not only membrane
proteins. Secondly, it is apparent that chloramphenicol had no *general* inhibitory effect on
protein synthesis (solid line). In particular, no
significant inhibition was observed for the in-

corporation of arginine-^{3}H into the fractions which
contain the major membrane polypeptides (from
9 to 11 cm from the origin), supporting the conclusion that these polypeptides are synthesized in
the cytoplasm (5). However, incorporation of
arginine-^{3}H into the fractions containing the large
(at 5.8 cm) and the small (at 14.2 cm) subunits of
ribulose 1,5-diphosphate carboxylase was inhibited. Thus, both subunits of this enzyme of
the chloroplast are apparently made on chloroplastic ribosomes.

Although electron microscopy indicated little
membrane formation in chloramphenicol treated
cells, the relative amounts of the membrane
polypeptides in membrane were determined as a
further test of the amount of the thylakoid membrane present in these cells. A total membrane
fraction was, therefore, prepared by centrifuging
the broken-cell preparations at 50,000 g_{av} for 15
min.[3] Electrophoresis of the protein of these
fractions produced the patterns of stain shown in
Fig. 7 A. Compared to the control, only small
amounts of the major polypeptides of thylakoid
membranes were present in the membrane
fraction from chloramphenicol-treated cells. In
the same experiment, the polypeptides were
labeled with arginine-^{3}H as described above for
the experiment shown in Fig. 6. Fig. 7 B shows the
radioactivity recovered in polypeptides in the
membrane fraction. (Since the top portion of the
gel contained little radioactivity and no significant
difference was observed between the two samples
in the region 0–3 cm, as in Fig. 6, the top portion
of the gels was not fractionated in this and subsequent experiments.) In control cells a large
amount of radioactive fractions *a*, *b*, and *c* was
incorporated into membranes. The pattern of
radioactivity for the control sample (dashed line)
was similar to that found previously for isolated
thylakoid membranes (5). Labeled subunits of
ribulose 1,5-diphosphate carboxylase were also
present, possibly sedimenting in association with
membranes (23). However, little radioactivity
was found in these fractions from chloramphenicol-
treated cells (solid line). In control cells, fraction

[3] Although this sample would be contaminated by
other cellular membranes, a total membrane fraction was prepared to enable a quantitative evaluation
of the amount of thylakoid membranes present. From
the electrophoretic patterns, it is apparent that
thylakoid membranes are the predominant constituent of this fraction in control cells.

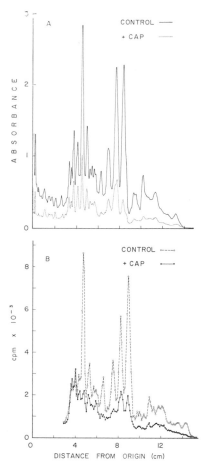

FIGURE 7 Patterns of protein stain (Fig. 7 A) and radioactivity (Fig. 7 B) after electrophoresis of polypeptides in the total membrane fraction from cells incubated with or without chloramphenicol (200 μg/ml). The experimental conditions were as described under Fig. 6. Arginine-³H was added at 5 hr to 1 μCi/ml, and 2 hr later the cells were collected, broken, and centrifuged for 15 min at 50,000 g_{av} at 2°C. The pellet fractions were prepared for electrophoresis as described under Methods, and the polypeptides were subjected to electrophoresis at 1.8 v/cm for 30 min and 6 v/cm for 6 hr. After staining, the gels were scanned at 563 mμ or sectioned for a determination of radioactivity as described under Fig. 6.

c contained more label than fraction b, but the labeling, albeit low, was reversed in chloramphenicol-treated cells. Thus, relatively more of fraction b had entered membrane material than did fraction c in the treated cells. The amounts of fraction c in membranes, estimated from results as shown in Fig. 7, were comparable to the amounts of chlorophyll present. In these experiments, chloramphenicol-treated cells contained 20–25% of the chlorophyll found in control cells (e.g., Fig. 5).

In control cells, newly synthesized polypeptides were continuously incorporated into growing membranes (Fig. 7 B). However, in chloramphenicol-treated cells fewer polypeptides were incorporated into thylakoid membranes. If in treated cells the membrane polypeptides continue to be synthesized in the cytoplasm at a rate higher than the rate of incorporation into thylakoid membranes, and if these polypeptides are synthesized in a soluble form and are in transit through the cell sap or chloroplast, then they should be detectable among the proteins of the soluble fraction under these experimental conditions. As a test of this assumption, the broken-cell preparations were centrifuged for 3 hr at 120,000 g_{av} so as to sediment particulate material. The protein in the supernatant fluid was then subjected to electrophoresis. Fig. 8 shows the patterns of protein stain (Fig. 8 A) and of radioactivity (Fig. 8 B) for the soluble polypeptides from control and chloramphenicol-treated cells. The control gel contained a small amount of radioactivity (dashed line) in the region of the major membrane polypeptides (8–10 cm from the origin). But in the sample from chloramphenicol-treated cells a much greater amount of radioactivity was found at 9.2 cm (arrow). This peak of radioactivity was in the position expected for membrane fraction c. Since there was no significant difference in this region in radioactivity from the total protein samples (Fig. 6), the results shown in Figs. 7 and 8 suggest that in the chloramphenicol-treated cells fraction c was accumulating in the soluble fraction instead of entering the membranes, as occurred in control cells. In one experiment, the difference in this region between the amounts of radioactivity in the samples of soluble protein from control and treated cells was about 3800 cpm. The difference in an equivalent amount of the membrane fraction was about 4600 cpm. Thus, approximately 80% of the amount missing from membranes in

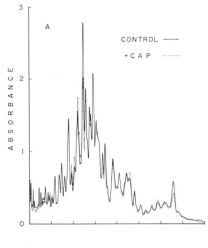

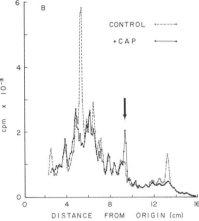

FIGURE 8 Patterns of protein stain (Fig. 8 A) and radioactivity (Fig. 8 B) after electrophoresis of soluble polypeptides from cells incubated with or without chloramphenicol (200 μg/ml). The experimental conditions were as described under Fig. 6. Arginine-³H was added at 5 hr to 1 μCi/ml, and 2 hr later the cells were collected, broken, and centrifuged for 3 hr at 120,000 g_{av} at 2°C. Protein in the supernatant fluids was prepared for electrophoresis as described in Methods. Portions of the samples containing 90 or 180 μg of the polypeptides were applied to companion gel columns and subjected to electrophoresis at 1.8 v/cm for 30 min and 6 v/cm for 6.5 hr. After staining, gels containing the smaller loads were scanned, while gels containing the higher loads were sectioned to determine the

chloramphenicol-treated cells was recovered in the soluble fraction. Only a small difference in the protein stain was found at this position in these short-term experiments (Fig. 8 A).

The amount of this polypeptide, characteristic of fraction c, in the soluble fraction was related to the degree of inhibition of membrane formation. When chloramphenicol was added after 3 hr of exposure to light, near the end of the slow phase of greening (Fig. 5), chlorophyll synthesis was inhibited during the time period of the experiments to a lesser extent than when the drug was added at 0 hr, as described above. In such experiments, after 7 hr of light exposure the treated cells contained 65 75% of the amount of chlorophyll contained in control cells. Arginine-³H was added at 5 hr, and 2 hr later the cells were collected and the soluble fractions were prepared. Electrophoresis of the soluble protein revealed only a slight increase, compared to Fig. 8, in radioactivity in the position of fraction c for treated cells over that for control cells. Also, only a correspondingly small decrease in radioactivity in this polypeptide was observed in the membrane fraction from such treated cells when compared to this fraction from control cells. Since less of this polypeptide fraction was recovered in the soluble fraction and more in the membrane fraction from cells in which membrane formation was inhibited less, these experiments provided further evidence that the soluble polypeptide is related to the membrane, and is fraction c of thylakoid membranes. Also, these results show that the appearance of the polypeptide in the soluble fraction was not simply the result of solubilization of a component of membranes assembled in the presence of the drug.

Attempts to chase the soluble protein into membrane after removing chloramphenicol were unsuccessful, since the cells recovered from the treatment only after a lag period of several hours, a period deemed too long to provide a clear result.

These experiments have not revealed the fate of membrane fractions a and b. Minor membrane polypeptides were present at concentrations too low to detect by the procedure used.

radioactivity pattern as described under Fig. 6. The chlorophyll level increased during the experiment from 0.7 μg/10⁷ cells at 0 hr to 1.2 μg and 6.8 μg of chlorophyll/10⁷ cells in chloramphenicol-treated and control cells, respectively.

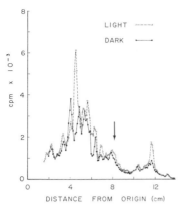

FIGURE 9 Electrophoretic analysis of the incorporation of arginine-³H into soluble polypeptides of greening *C. reinhardi* cells incubated 1 hr in light or dark. Yellow cells were suspended to 6 × 10⁶ cells/ml in fresh medium and exposed to light as described under Fig. 6, except that chloramphenicol was not added. After 6 hr of greening, one flask was wrapped with foil and placed in the dark. Arginine-³H was added to a concentration of 1 µCi/ml to both the flask in the dark and to an equivalent flask kept in the light. 1 hr later, the cells were collected and the soluble fractions were prepared for electrophoresis as described under Methods. Portions of the samples containing about 160 µg of the polypeptides were applied to companion gel columns and subjected to electrophoresis at 1.8 v/cm for 20 min and 6 v/cm for 6 hr. Radioactivity was determined as described in Fig. 6. Chlorophyll levels at the end of the experiment were: 12.4 µg/10⁷ cells (no increase during the labeling period) and 16.3 µg/10⁷ cells for cells placed in the dark and kept in light, respectively.

Effect of Light on Protein Synthesis

Since treating cells with chloramphenicol caused concomitant inhibition of chlorophyll synthesis and accumulation of the fraction *c* polypeptide in the soluble fraction of the cell, a relationship between this polypeptide and chlorophyll was suggested. In these cells, chlorophyll synthesis can also be stopped by turning off the light (3). Thus, a result similar to Fig. 8 might be expected if cells were placed in the dark instead of being treated with chloramphenicol. Fig. 9 shows the results obtained from an experiment in which cells, after 6 hr of greening, were placed in the dark. Arginine-³H was then added, and incorporation of the amino acid into soluble proteins by cells in the dark was compared after electrophoresis to that of an identical sample of cells kept in the light. No increase in radioactivity was observed for samples labeled in the dark in the region of the gel to which fraction *c* would migrate (arrow). Total membrane fractions (50,000 *g* pellet) were examined in order to determine whether labeled polypeptides were present in membranes. Fig. 10 shows, however, that a marked reduction of labeling of the membrane polypeptides occurred in cells transferred to the dark. Thus, apparently the synthesis of these major polypeptides as well as chlorophyll decreases in the dark. This agrees with results shown in Fig. 3, since after several days in the dark, the membrane polypeptides could not be observed in samples of total protein of yellow cells.

In confirmation of results reported recently by Schor et al. (24), no labeling of ribulose 1,5-diphosphate carboxylase occurred during the 1st hr after cells were transferred to the dark. Darkness was as effective as chloramphenicol in stopping synthesis of this enzyme. Lack of synthesis of other proteins was also detected (e.g., fractions near 6 cm from the origin in Fig. 9). Yet the synthesis of most soluble proteins appeared unaffected. These results provide further evidence that protein

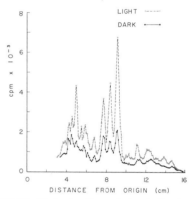

FIGURE 10 Electrophoretic analysis of the incorporation of arginine-³H into total membrane polypeptides of greening *C. reinhardi* cells incubated 1 hr in light or dark. Experimental conditions were as for Fig. 9. Membrane fractions were prepared as in Fig. 7, and the polypeptides were subjected to electrophoresis at 1.8 v/cm for 30 min and 6 v/cm for 6.5 hr. Radioactivity was determined as described under Fig. 6.

synthesis does not occur in the chloroplast immediately after cells are transferred to the dark. But the extent of protein synthesis in the cytoplasm, as evidenced by the extent of labeling of soluble proteins (Fig. 9), was considerably more than was expected from data reported by Ohad et al. (3) for the incorporation of labeled acetate into protein in a similar type of experiment.

DISCUSSION

The array of thylakoid membranes within the chloroplast is a prominent structural feature in *C. reinhardi* cells grown in light. These membranes contain two main polypeptide components which, at present, can only be identified by electrophoresis. Thus, to determine the fate of these polypeptides, it was necessary to establish that they could be detected after electrophoresis of total protein of the cells. The main membrane polypeptides are not prominent components of the cells, but Figs. 2 and 3 show that they can be detected.

Since chlorophyll is a constituent of the thylakoid membranes (25), determining the chlorophyll content during greening provided an approximation of the amount of membranes present. When cells were treated with a relatively high level of chloramphenicol (200 μg/ml), an inhibitor of protein synthesis in the chloroplast (17–20), chlorophyll synthesis and membrane formation were inhibited. But previous results, obtained at a lower chloramphenicol concentration (25 μg/ml), showed that membrane assembly continued in spite of extensive inhibition of synthesis of membrane proteins in the chloroplast. The marked decrease in membrane production at the higher concentration (200 μg/ml) was possibly the result of effects of chloramphenicol on the synthesis of membrane lipids or on the production of enzymes involved in the synthesis of chlorophyll.[2]

When chlorophyll synthesis was inhibited in the presence of chloramphenicol, polypeptides which are synthesized in the cytoplasm (5) and whose characteristics are those of fraction *c* polypeptides of thylakoid membranes appeared in the soluble fraction of the cell (Fig. 8). Conversely, when the synthesis of these polypeptides was inhibited by cycloheximide, an abrupt halt in the accumulation of chlorophyll and membranes resulted (4). Eytan and Ohad (6) showed, however, that prior treatment of *C. reinhardi* with chloramphenicol under conditions similar to those described in this paper relieved the inhibition of chlorophyll synthesis by cycloheximide. Pretreatment of *Euglena* cells with chloramphenicol also produced similar results (26). Thus, treatment with chloramphenicol apparently allowed the accumulation of proteins, made on cytoplasmic ribosomes, which are needed for chlorophyll accumulation. As Fig. 8 shows, the only soluble[4] polypeptide that accumulated in *C. reinhardi* during treatment with chloramphenicol was the fraction corresponding to membrane fraction *c*. Concomitantly, a decrease in the amount of fraction *c* in membranes was found in chloramphenicol-treated cells comparable to the decrease in the amount of chlorophyll in these cells. This evidence, albeit circumstantial, suggests a relationship between these two membrane components. Since the synthesis of chlorophyll appears confined to the chloroplast (27, 28), assembly of the membrane can, therefore, occur only in the chloroplast.

The fate of the other major polypeptide, fraction *b*, during treatment with chloramphenicol is less clear than that of fraction *c*. Relatively more of fraction *b* than of fraction *c* entered membrane material in the treated cells (Fig. 7), but this amount was still small compared to that in control cells. Yet, none of fraction *b* was observed in the soluble fraction of these cells. In the presence of chloramphenicol a decrease in the synthesis of this polypeptide possibly occurred.

Although fraction *c* appears to be an integral part of the thylakoid membranes, its fate when cells are treated with chloramphenicol resembles that of the inner mitochondrial membrane appendage, the coupling factor F_1, in yeast cells treated with this drug. F_1, also synthesized in the cytoplasm, accumulated as a soluble protein when mitochondrial development was inhibited (29). This suggested that some proteins do not attach to the membrane unless the proper sites have been prepared for them. A coupling factor from chloroplasts, CF_1, whose subunits have a molecular weight of 62,000, has been purified (30). This chloroplast coupling factor also appears to be synthesized in the chloroplast (31). It, therefore, is distinct from the major polypeptides shown in Fig. 1.

---

[4] It should be emphasized that in this context "soluble" refers to character and not location within the cell, since these experiments cannot indicate whether at the time of cell disruption the polypeptide was in the cytosol, in the chloroplast matrix, or in both.

On the basis of the effects of chloramphenicol, the site of synthesis in *C. reinhardi* of both subunits of ribulose 1,5-diphosphate carboxylase is the chloroplast. This conclusion was previously reached by Margulies (20, 32) and Smillie et al. (33) for bean and *Euglena* cells, respectively. Several reports have suggested the involvement of cytoplasmic ribosomes in the synthesis of this chloroplastic enzyme because of the inhibitory effects of cycloheximide on its synthesis (34, 35). However, there is some evidence that suggests the messenger RNA for this enzyme is made in the nucleus. Several mutations affecting this enzyme in tomato plants are located on nuclear chromosomes (36). Also, rifampicin, which inhibited chloroplast RNA synthesis in *C. reinhardi* (37), did not affect its synthesis (35). Rather than inhibiting translation of the messenger RNA for this enzyme, cycloheximide might interfere with the production or transfer of RNA from the nucleus to the chloroplast.

Results shown in Figs. 9 and 10 suggest a role for light in the regulation of protein synthesis in *C. reinhardi*. As observed by Schor et al. (24), protein synthesis stopped abruptly in the chloroplast when cells were placed in the dark. This would explain the absence of label in the subunits of the enzyme ribulose 1,5-diphosphate carboxylase. Nevertheless, activities in the chloroplast must resume within a short time since ribulose 1,5-diphosphate carboxylase (2) and chloroplastic ribosomes (18) and DNA (38) are synthesized when the cells remain in the dark.

The synthesis in the cytoplasm of the main polypeptides of the thylakoid membranes also apparently is regulated by light. Although in the dark the rate of synthesis of these polypeptides did not decrease as rapidly as that of ribulose 1,5-diphosphate carboxylase, as judged from the amount of arginine-^{3}H incorporated into these polypeptides (Fig. 10), synthesis of these components eventually decreased to a rate resulting in a level too low to detect in yellow cells. The mechanism of this regulation of protein synthesis by light is not known.

I thank Dr. George Palade for valuable discussions and for critically reading this manuscript. This investigation was supported by grant GB-8031 from the National Science Foundation.

REFERENCES

1. Sager, R., and G. E. Palade. 1954. *Exp. Cell Res.* **7**:584.
2. Ohad, I., P. Siekevitz, and G. E. Palade. 1967. *J. Cell Biol.* **35**:521.
3. Ohad, I., P. Siekevitz, and G. E. Palade. 1967. *J. Cell Biol.* **35**:553.
4. Hoober, J. K., P. Siekevitz, and G. E. Palade. 1969. *J. Biol. Chem.* **244**:2621.
5. Hoober, J. K. 1970. *J. Biol. Chem.* **245**:4327.
6. Eytan, G., and I. Ohad. 1970. *J. Biol. Chem.* **245**:4297.
7. Arnon, D. I. 1949. *Plant Physiol.* **24**:1.
8. Vernon, L. P. 1960. *Anal. Chem.* **32**:1144.
9. Lowry, O. H., N. J. Rosebrough, A. L. Farr, and R. J. Randall. 1951. *J. Biol. Chem.* **193**:265.
10. Loening, U. E. 1967. *Biochem. J.* **102**:251.
11. Yamashita, T., and T. Horio. 1968. *Plant Cell Physiol.* **9**:267.
12. McCarty, R. E., and E. Racker. 1966. *Brookhaven Symp. Biol.* **19**:202.
13. Johnson, U. G., and K. R. Porter. 1968. *J. Cell Biol.* **38**:403.
14. Rutner, A. C. 1970. *Biochem. Biophys. Res. Commun.* **39**:923.
15. Kawashima, N., and S. G. Wildman. 1970. *Biochem. Biophys. Res. Commun.* **41**:1463.
16. Kawashima, N., and S. G. Wildman. 1970. *Annu. Rev. Plant Physiol.* **21**:325.
17. Boulter, D. 1970. *Annu. Rev. Plant Physiol.* **21**:91.
18. Hoober, J. K., and G. Blobel. 1969. *J. Mol. Biol.* **41**:121.
19. Ellis, R. J. 1969. *Science (Washington).* **163**:477.
20. Margulies, M. M., and C. Brubaker. 1970. *Plant Physiol.* **45**:632.
21. Margulies, M. M. 1962. *Plant Physiol.* **37**:473.
22. Hudock, G. A., G. C. McLeod, J. Moravkova-Kiely, and R. P. Levine. 1964. *Plant Physiol.* **39**:898.
23. Howell, S. H., and E. N. Moudrianakis. 1967. *Proc. Nat. Acad. Sci. U.S.A.* **58**:1261.
24. Schor, S. L., P. Siekevitz, and G. E. Palade. 1970. *J. Cell Biol.* **47** (2, Pt. 2):182a. (Abstr.)
25. Walne, P. L., A. H. Haber, and L. L. Triplett. 1970. *Proc. Nat. Acad. Sci. U.S.A.* **67**:1501.
26. Smillie, R. M., N. S. Scott, and D. Graham. 1968. *In* Comparative Biochemistry and Biophysics of Photosynthesis. K. Shibata, A. Takamiya, A. T. Jagendorf, and R. C. Fuller, editors. University of Tokyo Press, Tokyo. 332.
27. Carell, E. F., and J. S. Kahn. 1964. *Arch. Biochem. Biophys.* **108**:1.
28. Rebeiz, C. A., and P. A. Castelfranco. 1971. *Plant Physiol.* **47**:24.
29. Tzagoloff, A. 1969. *J. Biol. Chem.* **244**:5027.
30. Farron, F. 1970. *Biochemistry.* **9**:3823.

31. RANALLETTI, M., A. GNANAM, and A. T. JAGEN-DORF. 1969. *Biochim. Biophys. Acta.* **186**:192.
32. MARGULIES, M. M. 1964. *Plant Physiol.* **39**:579.
33. SMILLIE, R. M., D. GRAHAM, M. R. DWYER, A. GRIEVE, and N. F. TOBIN. 1967. *Biochem. Biophys. Res. Commun.* **28**:604.
34. CRIDDLE, R. S., B. DUN, G. E. KLEINKOPF, and R. C. HUFFAKER. 1970. *Biochem. Biophys. Res. Commun.* **41**:621.

35. ARMSTRONG, J. J., S. J. SURZYCKI, B. MOLL, and R. P. LEVINE. 1971. *Biochemistry.* **10**:692.
36. ANDERSEN, W. R., G. F. WILDNER, and R. S. CRIDDLE. 1970. *Arch. Biochem. Biophys.* **137**:84.
37. SURZYCKI, S. J. 1969. *Proc. Nat. Acad. Sci. U.S.A.* **63**:1327.
38. CHUN, E. H. L., M. H. VAUGHAN, and A. RICH. 1963. *J. Mol. Biol.* **7**:130.

Cytoplasmic and Chloroplast Ribosomes

ISOLATION OF STABLE RIBOSOMAL RNA FROM WHOLE CELLS OF CHLAMYDOMONAS REINHARDTII

ROSE ANN CATTOLICO AND RAYMOND F. JONES

SUMMARY

Sonication, rather than lysis, of cells in the presence of diethylpyrocarbonate and Mg^{2+} as nuclease inhibitors has resulted in the isolation of stable high molecular weight ribosomal RNA from *Chlamydomonas reinhardtii*. Preparations are nuclease free and show excellent conservation of the previously thought labile 23-S rRNA species.

INTRODUCTION

Total loss or partial degradation of ribosomal RNA species during RNA isolation in algae has been reported for *Acetabularia mediterranea*[1], *Chlamydomonas reinhardtii*[2,3], *Euglena gracilis*[4,5], and *Volvox carteri*[6]. Exploitation of these simple eukaryotic algal systems for the study of ribosomal RNA synthesis and maturation during growth and differentiation has, therefore, been impossible. In this paper, we report on an isolation procedure which has proven most effective in obtaining stable, nuclease-free preparations of high molecular weight RNA from whole cells of *Chlamydomonas reinhardtii*.

MATERIALS AND METHODS

Synchronous cultures of *Chlamydomonas reinhardtii* Dangeard strain 137C (*plus* mating strain) were grown on high salt minimal medium[7] as previously described[8].

For the isolation of RNA, cells of *Chlamydomonas* were harvested by centrifugation at 7000 rev./min for 20 min at 5°C using an RC2B Sorvall centrifuge and GSA rotor. 6 vol. of ice-cold buffer, pH 7.6, containing 0.025 M Tris, 0.025 M $MgCl_2$ and 0.025 M KCl was saturated, unless otherwise stated, with diethylpyrocarbonate and added to the pellet. In studies where the nuclease inhibitor Bentonite was tested, it was prepared as described by Nisman[9] and added to the Tris–$MgCl_2$–KCl buffer in concentrations ranging from 2–30 mg/ml instead of the diethylpyrocarbonate. The resuspended cells were then treated by one of the following methods:

(a) *Sonication.* Cells were sonicated for 30–45 s in an MSE ultrasonic disinte-grator at 5°C and cell breakage checked by phase-contrast microscopy. Sodium dodecylsulfate in Tris–MgCl$_2$–KCl buffer was added to produce a final concentration of 0.8%. After 2 min of occasional gentle stirring at 5°C, 4 vol. of phenol solution (redistilled crystalline phenol extracted with Tris–MgCl$_2$–KCl buffer *plus* diethyl-pyrocarbonate until neutral) was added. The suspension was stirred gently and after 5 min frozen at −70°C.

(b) *Lysis.* Sodium dodecylsulfate was added to produce a final concentration of 0.8%. After 2 min at 5°C, 4 vol. of phenol (prepared as above) were added, and the suspension stirred gently. After a further 5 min the suspension was frozen at −70°C.

Prior to RNA extraction the frozen suspensions were allowed to thaw in a water bath at 15°C, then shaken on a New Brunswick G2 Laboratory Rotor at 200 rev./min at 15°C. Each RNA preparation was centrifuged at 7000 rev./min in a Sorvall HB-4 rotor for 20 min at 15°C. The supernatant was collected, and extracted twice more with an equal volume of Tris–MgCl$_2$–KCl *plus* diethylpyrocarbonate-neutralized phenol as described above. The final supernatant was mixed and ex-tracted with ice-cold ether (15 times volume). Upon removing the residual ether by bubbling with nitrogen, 95 % ethanol (2.2 ml ethanol per ml supernatant) and 20 % potassium acetate (0.1 ml/ml supernatant) were added and the final preparation allowed to stand for a minimum of 18 h at −20°C. The precipated RNA was collected by cen-trifugation at 10 000 rev./min at 5°C using a Sorvall SS34 rotor, and the pellet was brought almost to dryness with nitrogen. A buffer, pH 7.5 containing 2 mM EDTA, 0.01 M NaCl, 0.1 M Tris saturated with diethylpyrocarbonate was added to give an RNA concentration of approx. 4 mg/ml. After 45 min of occasional stirring, the prep-aration was finally centrifuged at 10 000 rev./min in a Sorvall SS34 rotor at 5°C for 20 min, and the supernatant stored frozen in a glass tube. The RNA obtained by this method routinely produced a 260 nm/280 nm absorbance ratio of between 1.90 and 2.1.

The individual RNA species were separated and analysed by a modification of the polyacrylamide-gel electrophoresis technique of Peacock and Dingman[10]. Gels of 2.4 % acrylamide were routinely run at 5°C in a buffer, pH 7.9 containing 0.036 M Tris, 0.03 M NaH$_2$PO$_4$, and 1 mM EDTA[2], and finally scanned at 260 nm in a Gilford Spectrophotometer with a linear transport attachment.

RESULTS

As seen in Fig. 1A, 4 clearly separated rRNA species are obtained when cells of *C. reinhardtii* are sonicated in the high salt, high Mg^{2+} buffer containing diethyl-pyrocarbonate, prior to phenol addition. Although direct lysis of cells in phenol has been used as a standard technique in RNA extraction[11,12] it is seen (Fig. 1B) to be highly unsuitable for the preparation of RNA from *Chlamydomonas* for virtually all the *Chlamydomonas* rRNA is lost and there is an accompanied increase in lower molecular weight RNAs.

When a preparation of *Chlamydomonas* RNA was incubated, in the absence of diethylpyrocarbonate, at 5°C and 32°C before gel analysis, no difference in RNA species distribution was obtained (Figs 2A and 2B). These results indicate that the level of nuclease contamination in our preparations is not significant.

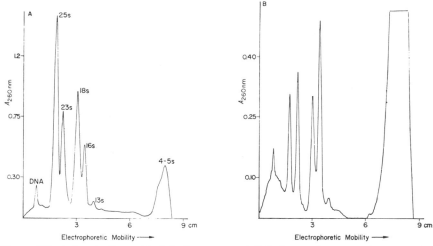

Fig. 1. Effect of sonication and lysis in the presence of diethylpyrocarbonate on the preparation of RNA from whole cells of *C. reinhardtii*. (A) Sonication, (B) Lysis. Electrophoresis was run for 2 h at 5°C.

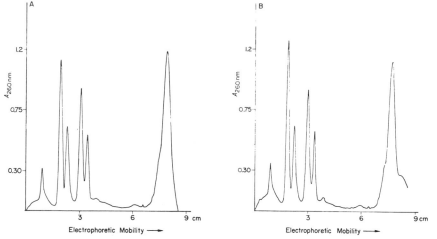

Fig. 2. Absence of ribonuclease contamination in RNA preparations from *C. reinhardtii*. (A) Incubation of RNA for 10 min at 5°C prior to electrophoresis at 5°C. (B) RNA incubated for 10 min at 32°C prior to electrophoresis at 5°C. Electrophoresis was run for 2 h.

rRNA distribution can be quite readily affected by slight variations in our reported isolation or analytical procedures. For example, the presence of the nuclease inhibitor Bentonite (2–30 mg/ml) during the sonication process, instead of diethyl-pyrocarbonate afforded little protection, but the RNA bands tended to be broader

TABLE I

EFFECT OF BENTONITE ON rRNA EXTRACTION

Percent distribution was obtained by calculating areas under the curves of gel scans.

Treatment	% Distribution of rRNA species	
	25-S+23-S	18-S+16-S
Sonication with saturated diethylpyrocarbonate	62.48±0.87	37.51±0.85
Sonication with Bentonite		
2 mg/ml	59.67±0.28	40.35±0.26
10 mg/ml	57.24±0.38	42.77±0.37
30 mg/ml	53.95±0.30	46.05±0.33

and a loss of the rRNA species was accompanied by an increase in degradation products. As seen in Table I, a loss in both large subunit rRNAs and a concomitant increase in the small subunit rRNA species also occurs. Bentonite added after sonication affords no protection.

Several recent studies on rRNA species in algae[1,5,6] and in higher plants[13,14] have been carried out using gel electrophoresis at room temperature. In our studies, we note that for *Chlamydomonas*, at least, such high temperatures are unsuitable. As can be seen from Fig. 3, a loss of the 23-S RNA band and a broadening of the remaiing RNA bands occurs when gels are run at 22°C.

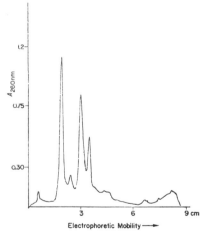

Fig. 3. Effect of temperature upon stability of *C. reinhardtii* RNA during electrophoresis. Electrophoresis was run for 2 h at 22°C.

DISCUSSION

The method of cellular breakage appears to be critical in the preparation of high-molecular weight rRNA from *C. reinhardtii*. Immediate disruption of the cells by sonication results in preparations with 4 clearly defined ribosomal species. Lysis, however, results in an almost total loss of rRNA accompanied by alteration in the distribution of the ribosomal species and a large increase in degredation products. It is interesting to note that the controversy[15,16] surrounding the possible absence of 23-S RNA in *Rhodopseudomonas speroides* was resolved when Szilagyi[17] demonstrated that disruption of the bacterium by sonication restored the 23-S component.

Diethylpyrocarbonate has proved to be very effective as a nuclease inhibitor in our isolation technique. The action of this newly discovered inhibitor[18] is still under investigation[19,20] but Wolf *et al.*[21] suggest that diethylpyrocarbonate causes the formation of intramolecular peptide-like bonds within the nuclease protein, which results in conformational changes and an irreversible loss of enzymatic activity.

The presence of Mg^{2+} has been reported to increase the stability of radish cotyledon RNA during isolation, and to hold "nicked" rRNA pieces together during gel analysis[22]. It should be noted that our isolation procedure for *Chlamydomonas* RNA contains Mg^{2+} and that when electrophoresis is carried out either in a Mg^{2+} or EDTA buffer system we have not observed any difference in RNA gel profiles. This fact, along with the inability to detect nuclease contamination in our preparations, leads us to believe that the isolated *Chlamydomonas* RNA species are intact.

Leaver and Ingle[23] have attributed the decreased yield in Swiss chard 23-S rRNA to the fact that Bentonite was suspended in EDTA which they showed to be harmful. We have found that the use of Bentonite results in the broadening of the rRNA bands in *Chlamydomonas*, and alteration in the distribution of rRNA species. Since our Bentonite was exhaustively washed with a buffer containing 25 mM Mg^{2+}, we feel that in our system, sonication with diatomatious earth may cause a sheer effect.

In plants the 23-S rRNA of chloroplasts has been reported to be quite labile. For example, 29–57 % of this RNA was lost in cucumber preparations[13], and an almost 100 % loss has been reported in *Volvox*[6], *Acetabularia*[1], and *Chlamydomonas*[23]. Our preparations, however, show excellent conservation of this ribosomal species.

The fact that, in this investigation, the RNA preparations were obtained from whole cells and not isolated ribosomes, renders the technique useful for ribosomal precursor maturation studies.

ACKNOWLEDGMENTS

This investigation was supported by Contract AT (30-1) 3475 with the U. S. Atomic Energy Commission.

REFERENCES

1 C. L. F. Woodcock and L. Bogorad, *Biochim. Biophys. Acta*, 224 (1970) 639.
2 U. E. Loening, *J. Mol. Biol.*, 38 (1968) 355.
3 D. P. Bourque, J. E. Boynton and N. W. Gillham, *J. Cell Sci.*, 8 (1971) 153.
4 J. R. Rawson, E. J. Crouse and E. Stutz, *Biochim. Biophys. Acta*, 246 (1971) 507.
5 K. E. Schuit, N. G. Avadhani and D. E. Buetow, *Arch. Mikrobiol.*, 71 (1970) 79.
6 G. Kochert and W. Sansing, *Biochim. Biophys. Acta*, 238 (1971) 397.
7 N. Sueoka and K. S. Chiang, *J. Mol. Biol.*, 25 (1967) 47.
8 J. R. Kates, Ph.D. Thesis, Princeton University, 1966.

9 B. Nisman, in L. Grossman and K. Moldave, *Methods in Enzymology*, Vol. 12B, Academic Press, New York, 1968, p. 794.
10 A. C. Peacock and C. W. Dingman, *Biochemistry*, 7 (1968) 668.
11 A. Trewavas, *Plant Physiol.*, 45 (1970) 742.
12 U. E. Loening and J. Ingle, *Nature*, 215 (1967) 363.
13 F. Vedel and M. J. d'Aoust, *Plant Physiol.*, 46 (1970) 81.
14 M. Edelman, I. M. Verma, R. Herzog, E. Galun and U. Z. Littauer, *Eur. J. Biochem.*, 19 (1971) 372.
15 B. Marrs and S. Kaplan, *J. Mol. Biol.*, 49 (1970) 297.
16 T. G. Lessie, *J. Gen. Microbiol.*, 39 (1965) 311.
17 J. F. Szilagyi, *Biochem. J.*, 109 (1968) 191.
18 F. Solymosy I. Fedorcsák, A. Gulyás, G. L. Farkas and L. Ehenberg, *Eur. J. Biochem.*, 5 (1968) 520.
19 C. G. Rosen and I. Fedorcsák, *Biochim. Biophys. Acta*, 130 (1966) 401.
20 D. G. Humm, J. H. Humm, and L. I. Shoe, *Biochim. Biophys. Acta*, 246 (1971) 458.
21 B. Wolf, J. A. Lesnaw and M. E. Reichmann, *Eur. J. Biochem.*, 13 (1970) 519.
22 J. Ingle, J. V. Possingham, R. Wells, C. J. Leaver and U. E. Loening, *Symp. Soc. Exp. Biol.*, 24 (1970) 303.
23 C. J. Leaver and J. Ingle, *Biochem. J.*, 123 (1971) 235.

CHLOROPLAST RIBOSOME DEFICIENT MUTANTS IN THE GREEN ALGA *CHLAMYDOMONAS REINHARDI* AND THE QUESTION OF CHLOROPLAST RIBOSOME FUNCTION

J. E. BOYNTON, N. W. GILLHAM AND J. F. CHABOT

SUMMARY

The 2 chloroplast ribosome deficient mutants of *Chlamydomonas reinhardi*, *ac-20* and *cr-1* form low levels of 66-s and 70-s chloroplast ribosome monomers compared to wild type, but differ in that *cr-1* also accumulates substantial amounts of the large (54-s) chloroplast subunit. Unlike wild type, *ac-20*, *cr-1* and the double mutant *ac-20cr-1* make little or none of the CO_2-fixing enzyme ribulose diphosphate carboxylase (RuDPCase) when grown in the light with acetate and CO_2 as carbon sources (*mixotrophic growth*) whereas all 3 mutant genotypes form at least some enzyme when grown photosynthetically with CO_2 as the sole carbon source (*phototrophic growth*). All 3 mutant genotypes also show a characteristic defect in the organization of their chloroplast lamellar system when grown mixotrophically, but their chloroplast membrane organization approaches that of wild type when grown phototrophically. No change in the level of chloroplast ribosomes accompanies these changes in chloroplast organization and RuDPCase formation. We propose that the pleiotropic effects on chloroplast structure and function in mixotrophically grown cells of the mutants result because their chloroplast ribosomes synthesize certain specific chloroplast components with low efficiency, whereas they perform this function with greater efficiency when the cells are grown phototrophically.

INTRODUCTION

In this paper we compare the ability of 2 chloroplast ribosome deficient mutants in *C. reinhardi*, *ac-20* and *cr-1*, to form chloroplast ribosomes, to synthesize the CO_2-fixing enzyme ribulose-1,5-diphosphate carboxylase (RuDPCase, EC 4.1.1.39) and to form a structurally normal chloroplast. The *ac-20* mutant is deficient in chloroplast ribosomes and several chloroplast components including RuDPCase when grown in the light with both CO_2 and acetate as carbon sources (*mixotrophic growth*) (Goodenough & Levine, 1970; Levine & Paszewski, 1970; Togasaki & Levine, 1970). We have described elsewhere the origin and chloroplast ribosome phenotype of *cr-1* and the double mutant *ac-20cr-1* (Bourque, Boynton & Gillham, 1971; Boynton, Gillham & Burkholder, 1970). The deficiency in chloroplast ribosomes in both *ac-20* and *cr-1* is correlated with a reduced level of the enzyme RuDPCase and with a characteristically abnormal organization of the chloroplast under mixotrophic growth conditions. When *ac-20*, *cr-1* or the double mutant *ac-20cr-1* are grown photosynthetically with only CO_2 as a carbon source (*phototrophic growth*) both RuDPCase level and chloroplast

89

organization approach wild type, but their chloroplast ribosome deficiencies remain unchanged. This led us to hypothesize that chloroplast ribosomes function more efficiently in making certain chloroplast protein components under phototrophic conditions than under mixotrophic conditions.

MATERIALS AND METHODS

Organisms and culture conditions

The wild type (*ac-20⁺cr-1⁺*), the double mutant (*ac-20cr-1*), and the 2 single mutant (*ac-20cr-1⁺* and *ac-20⁺cr-1*) clones used in these experiments were derived from one tetratype tetrad (5·6) resulting from a cross between an *ac-20⁺cr-1⁺* mating type plus (*mt⁺*) stock and an *ac-20cr-1* mating type minus (*mt⁻*) stock, described previously (Bourque *et al.* 1971; Boynton *et al.* 1970).

All genotypes were grown at 25 °C in 300-ml shake cultures to late logarithmic phase (3–5 × 10⁶ cells per ml) on the high salt medium of Sueoka (1960) with (HSA) or without (HS) the addition of 2 g/l. of sodium acetate as follows: *Mixotrophic growth*, in HSA medium under cool white fluorescent lights (~ 3500 lux); *heterotrophic growth*, in HSA medium in the dark in flasks covered with black insulating tape; *phototrophic growth*, in HS medium under cool white fluorescent lights (~ 3500 lux) and aerated with 5 % carbon dioxide in air mixture.

Genetic methods

Standard procedures were used for making crosses and for dissection and analysis of tetrads (Ebersold & Levine, 1959; Gillham, 1965).

Isolation and characterization of ribosomes on linear sucrose gradients

Ribosomes were prepared as described previously (Bourque *et al.* 1971), and centrifuged in 10–30% linear sucrose gradients in the SW 27 rotor at 22 500 rev/min for 14 h at 2 °C. Gradients were analysed for absorbance at 254 nm with a UA-2 UV analyser-ISCO model D gradient fractionator and the relative amounts of chloroplast and cytoplasmic ribosomes were then estimated (Bourque *et al.* 1971).

Comparisons of cytological organization and relative amounts of chloroplast and cytoplasmic ribosomes by electron microscopy

All procedures ranging from fixation of the cells to analysis of the electron micrographs were the same as those used previously by us (Bourque *et al.* 1971) with 2 exceptions. First, ribosome number was estimated by counting the number of electron-dense particles of ribosome size in 50 0·5-cm grid squares (0·7 μm²) in unobstructed areas of the cytoplasm and the chloroplast on 2 separate pictures of × 20000 final magnification of each of 10 cells. Secondly, the volume of a hypothetical cell and the number of chloroplast and cytoplasmic ribosomes per hypothetical cell were calculated for each cell analysed with the aid of a computer program and mean values determined.

Ribosome estimates in Table 4 (p. 276) for the genotypes *ac-20⁺cr-1⁺* grown mixotrophically and *ac-20cr-1* grown both mixotrophically and phototrophically reveal a substantially higher number of chloroplast and cytoplasmic ribosomes estimated per μm² area and extrapolated per hypothetical cell than observed previously (Bourque *et al.* 1971). However, the chloroplast ribosomes constitute roughly the same percentage of the total cell ribosomes in each case and the absolute differences may reflect the fact that different people made the counts or that the ribosomes were better visualized in recent preparations.

Chlorophyll determinations

Amounts of chlorophyll were determined by *in vivo* spectrophotometry (Bourque *et al.* 1971). Total chlorophyll was estimated by the following modification of Arnon's (1949) equation: μg chlorophyll/l. $= 20 \cdot 0 \; \text{O.D.}_{650} + 11 \cdot 5 \; \text{O.D.}_{678}$, where the extinction coefficients have been changed to correct for the differences between *in vivo* absorption at 650 nm and 678 nm and *in vitro* absorption of acetone extracts of the same amount of chlorophyll at 645 nm and 663 nm (DeVitis, unpublished).

Assay of RuDPCase

RuDPCase activity was measured by incorporation of $\text{NaH}^{14}\text{CO}_3$ into acid-stable products in the presence of ribulose-1,5-diphosphate (RuDP). The method used was derived from those described by Björkman (1968), Björkman & Gauhl (1969) and Levine & Togasaki (1965). A 100-μl aliquot of enzyme was added to 900 μl of reaction mixture in a 13-mm tube containing: 50 μM Tris-HCl, pH 7·5; 40 μM KHCO_3; $\text{NaH}^{14}\text{CO}_3$ (Calatomic) to give a specific activity for $\text{H}^{14}\text{CO}_3{}^-$ of 0·2 μCi/μM; 0·1 μM RuDP (Sigma); 2·5 μM MgCl_2; 1 μM dithiothreitol. The foregoing substrate concentrations for RuDP and KHCO_3 were established as optimal under our conditions by varying the concentration of each substrate independently for a given amount of enzyme. Reaction rates were linear for more than 30 min. Enzyme from every sample was assayed at several different concentrations to ensure that data would be used only from reaction tubes where substrates were not limiting.

The reaction was carried out in a water bath at 25 °C for 20 min and was then stopped by the addition of 0·3 ml of 6 M acetic acid. Reaction mixtures were poured into liquid scintillation vials and incubated at 80–100 °C to drive off unfixed $^{14}\text{CO}_2$ and reduce the remaining material to dryness. A 0·1-ml aliquot of water was added to each vial to redissolve the residue, followed by 10 ml of Bray's solution (Bray, 1960). The samples were counted in a Packard 314 liquid scintillation spectrometer, corrected for background and then for quench using a channels ratio method. Michaelis constants of the highly purified enzyme for the substrates RuDP and $\text{HCO}_3{}^-$ were plotted by the method of Lineweaver & Burke (1934) and the line fitted by the least squares method.

Purification of the enzyme RuDPCase

For preparation of crude extracts or purification of RuDPCase, cells were harvested and then broken in Buffer A of Hoober & Blobel (1969) as described previously (Bourque *et al.* 1971). When both crude extracts and purified enzyme were to be analysed in the same sample, the homogenate was first centrifuged at 12 000 *g* for 10 min and then the supernatant decanted and separated into 2 aliquots. One was loaded directly on 10–30 % linear sucrose gradients for enzyme purification while the other was centrifuged at 27 000 *g* for 20 min and the assay of the crude extract then performed on the supernatant of this centrifugation. Appropriate enzyme dilutions were made using Buffer A. All steps were carried out at 4 °C unless otherwise specified. Partial purification of RuDPCase was achieved in the same linear sucrose gradients used to separate and estimate chloroplast and cytoplasmic ribosomes. Under these conditions of centrifugation the cytoplasmic ribosomes move about 90 % of the length of the gradient and RuDPCase appears as a distinct shoulder (absorbance at 254 nm) in the 18–21 s region of the gradient partially separated from the nucleic acids and proteins near the top (Fig. 1). In a preliminary experiment with our wild type stock, RuDPCase activity was confined *exclusively* to this 18–21 s region of the gradient (Fig. 1).

Highly purified RuDPCase was obtained by centrifuging the sucrose gradients for 17 or 24 h at 25 000 rev/min at 5 °C in the SW 27 rotor and using absorbance at 254 nm to locate the peak. Under these conditions the cytoplasmic and chloroplast ribosome monomers are pelleted and the RuDPCase peak separates almost completely from the proteins and nucleic acids at the top of the gradient (Fig. 3, p. 279). The ratio of the optical densities at 280 and 260 nm of the enzyme peak of wild type in these preparations was 1·63, indicating that nucleic acid contamination of the enzyme peak was less than 0·3 %. Therefore the absorbance measured at 254 nm was caused by the RuDPCase protein and not contaminating nucleic acids. The amount of purified RuDPCase was estimated by extrapolating both the RuDPCase peak and the nucleic acid-protein peak at the top of the gradient to base-line after which the area under the RuDPCase peak not overlapped by the peak at the top of the gradient was cut out and weighed. This value was then corrected to mg protein as estimated by the Lowry protein method (see below).

One-millilitre fractions were routinely collected for enzyme assay from gradients. In experiments involving the highly purified enzyme, the 3 or 4 peak tubes, as indicated by absorbance at 254 nm, were usually pooled prior to assay, whereas in experiments involving partially purified enzyme the individual fractions were assayed separately (Fig. 2, p. 274) and the data subsequently pooled for tabulation (Table 7, p. 280). Aliquots of all preparations assayed for

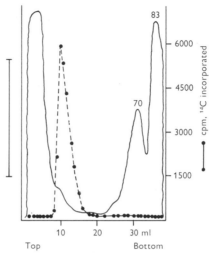

Fig. 1. Sucrose gradient profile of chloroplast and cytoplasmic ribosomes and the enzyme RuDPCase from mixotrophically grown wild type cells of *Chlamydomonas*. ——, absorbance at 254 nm (bar = 1·0 O.D. unit); – – –, H^{14}CO$_3^-$ incorporated into acid-insoluble products. The large activity peak in the 18–21 s region of the gradient represents RuDPCase. The presence of this enzyme is also evident as a small shoulder in the absorbance at 254 nm.

RuDPCase activity were frozen and stored at −20 °C until protein could be measured. Upon thawing the protein was precipitated with 20 % trichloroacetic acid for 20 min at room temperature prior to being pelleted by centrifugation at 27 000 g for 20 min. The protein pellet was then redissolved in 0·2 N NaOH and assayed by the method of Lowry, Rosebrough, Farr & Randall (1951). Bovine Serum Albumin (Sigma) was used as a standard. In a reconstruction experiment using known amounts of Bovine Serum Albumin, recovery of protein by this method exceeded 90 % down to concentrations as low as 0·01 mg protein/ml.

Goldthwaite & Bogorad (1971) recently published a paper describing virtually the same method as ours for purification of RuDPCase and found the enzyme peak detected on sucrose gradient centrifugation to be homogenous on acrylamide gel electrophoresis. It is reasonable to assume that our highly purified enzyme preparations from wild type and at least one of our mutants would show equally little contamination with extraneous protein.

Genetics of the ac-20 and cr-1 mutants

Both *ac-20* and *cr-1* are Mendelian mutations which affect chloroplast ribosome phenotype in different ways (fig. 2, table 3 of Boynton *et al.* 1970). The *ac-20* mutant maps in linkage group XIII (Hastings *et al.* 1965) and *cr-1* is not linked to *ac-20*

Table 1. *Genetic segregation of the mutants ac-20 and cr-1 observed in crosses of the double mutant ac-20 cr-1 with wild type ac-20⁺ cr-1⁺*

	Segregations observed							
Crosses*	Acetate requirement†			No. scored	Chloroplast ribosome phenotype‡			No. scored
	PD	NPD	T		PD	NPD	T	
ac-20^+ cr-1^+ mt^+ × ac-20 cr-1 mt^-	7	10	28	45	1	3	4	8
ac-20 cr-1 mt^+ × ac-20^+ cr-1^+ mt^-	3	1	4	8	—	—	—	—
Totals	10	11	32	53	1	3	4	8

* Crosses summarized include data published previously by Boynton *et al.* (1970).

† Acetate requirement segregation: PD, 2:2 acetate dependent:independent; NPD, 4:0 acetate dependent:independent; T, 3:1 acetate dependent:independent.

‡ Since only selected tetrads were examined for ribosome phenotype, the ratios should be ignored for linkage determination. Ribosomal phenotype segregation: PD = 2 *ac-20 cr-1* : 2 *ac-20⁺ cr-1⁺*, NPD = 2 *ac-20 cr-1⁺* : 2 *ac-20⁺ cr-1*, T = 1 *ac-20⁺ cr-1⁺* : 1 *ac-20 cr-1* : 1 *ac-20 cr-1⁺* : 1 *ac-20⁺ cr-1*. Ribosome profiles of each of these four genotypes are shown in Fig. 2.

(Table 1). If the 2 mutations were linked the number of parental ditype tetrads (PD) would greatly exceed the number of non-parental ditype tetrads (NPD) because the NPD tetrads can only arise from a 4-strand double crossover between the 2 genes in question. In the case of unlinked genes the frequencies of PD and NPD tetrads should be equal since both kinds of tetrads arise as a result of independent assortment. Tetratype (T) tetrads in both cases arise from crossovers between the mutant genes and their centromeres.

When *cr-1* is crossed to wild type the expected 2:2 Mendelian segregation for acetate requirement (5 tetrads analysed) and chloroplast ribosome phenotype (3 tetrads analysed) is observed.

All experiments reported in this paper involve the 4 genotypes of a single tetratype tetrad, designated 5·6, from a cross between wild type (*ac-20⁺ cr-1⁺*) and the double mutant (*ac-20 cr-1*) and each genotype was verified as described previously (Boynton *et al.* 1970). Hereafter the genotype *ac-20⁺ cr-1⁺* (product 5·6·4) will be called wild type; the genotype ac-20^+cr-1 (product 5·6·3), cr-1; the genotype ac-$20cr$-1^+ (product 5·6·1), ac-20; and the genotype ac-$20cr$-1 (product 5·6·2) the double mutant.

Comparison of growth rates

Growth rates of all 4 genotypes were measured for cells grown phototrophically, mixotrophically and heterotrophically. With the exception of ac-20 grown heterotrophically, cells of all 3 of the mutant genotypes had doubling times during

Table 2. *Growth parameters of the 4 genotypes of the tetratype tetrad 5·6 from the cross* ac-20^+cr-$1^+mt^+ \times ac$-$20cr$-$1mt^-$. *Mean values are given for phototrophic (P), hetero-trophic (H) and mixotrophic (M) growth conditions.*

Genotype	Growth conditions	Cell doubling time, h	P/M	H/M	P/H	Mutant/ wild type	Final cell concentration × 10⁶	No. of determi- nations
ac-$20^+ cr$-1^+	P	12·8				—	4·16	2
	H	14·9	1·3	1·5	0·9	—	4·05	1
	M	9·7				—	8·58	2
ac-$20 cr$-1	P	19·9				1·6	3·17	2
	H	17·6	1·3	1·2	1·1	1·2	3·15	1
	M	14·9				1·5	4·73	2
ac-$20 cr$-1^+	P	18·9				1·5	4·16	2
	H	13·6	1·5	1·1	1·4	0·9	5·55	1
	M	12·7				1·3	4·53	2
ac-$20^+ cr$-1	P	21·1				1·6	2·29	2
	H	18·4	1·3	1·1	1·1	1·2	3·12	1
	M	16·8				1·7	5·45	1

logarithmic phase longer than wild type irrespective of growth conditions (Table 2). Relative growth rates for ac-20, cr-1, and the double mutant under the 3 sets of growth conditions expressed as the ratio of doubling times under different pairs of growth conditions were very similar to those of wild type (Table 2). The ratio for phototrophically to mixotrophically grown cells ranges from 1·3 to 1·5, indicating that cells of all 4 genotypes grow more rapidly in the light when acetate is added to the medium, but that none of the mutants show a proportionately greater stimulation of growth by acetate in the light than does wild type. The ratio of doubling times for heterotrophically to mixotrophically grown cells is 1·5 for wild type and ranges from 1·1 to 1·2 for the 3 mutant genotypes, indicating that wild type grows somewhat better than the mutants on acetate-containing media in the light than in the dark. The ratio of doubling times for heterotrophically to phototrophically grown cells of cr-1, the double mutant and wild type, is close to 1·0, indicating that growth on CO_2 as a carbon source in the light and growth on acetate in the dark are about equally efficient for all 3 genotypes. The ratio of 1·4 for ac-20 suggests that acetate-stimulated growth in the dark may be somewhat more efficient than photosynthetic growth. Most significantly, none of the 3 mutant genotypes are stimulated in terms of growth rate to a proportionately greater extent than wild type by the presence of acetate in the medium. That is, ac-20 and cr-1 do not behave like most of the other acetate-requiring mutants known in C. reinhardi (Levine & Goodenough, 1970).

Why then can the 3 mutant genotypes be differentiated from wild type by the fact that they grow less well on agar plates of minimal medium than on plates of acetate-containing medium ? The explanation may be as follows. Tetrads for genetic analysis are dissected on agar plates of acetate-containing medium and grown up in the light (i.e. mixotrophically). Wild-type cells grown mixotrophically in liquid medium have as much RuDPCase as phototrophically grown cells whereas the 3 mutant genotypes have practically no enzyme when grown mixotrophically, although all 3 genotypes make the enzyme when grown phototrophically. Thus when tetrads grown mixotrophically are transferred by replica-plating to minimal medium lacking acetate, the wild-type clones can begin to grow immediately by fixing CO_2, but the mutant clones will undergo a growth lag the length of which will depend upon how long the cells take to synthesize enough RuDPCase to begin fixing CO_2. Therefore the mutant genotypes appear as acetate requirers initially upon transfer. Since the mutant genotypes are ultimately capable of phototrophic growth on solid media, such plates must be scored soon after replica-plating or it becomes difficult to distinguish the mutants from wild type.

Ribosome phenotypes of the 4 genotypes

Mixotrophically grown cells of ac-20 and the double mutant make fewer chloroplast ribosome monomers than wild type whereas cr-1 makes about the same amount of chloroplast ribosome monomers as the other 2 mutant genotypes but in addition preferentially accumulates the large (54-s) subunit of the chloroplast ribosome (Boynton et al. 1970; Bourque et al. 1971). We have extended these comparisons to the 4 genotypes of the 5·6 tetrad grown both phototrophically and mixotrophically and shown that while the ribosome phenotypes of the 4 genotypes differ from one another, they remain unchanged for a given genotype under either growth condition (Fig. 2, Table 3). Wild-type cells have a large peak of 70-s chloroplast ribosome monomers and often a smaller peak of 66-s monomers in addition to the 83-s cytoplasmic ribosome peak. The 2 other minor peaks represent chloroplast ribosome subunits the larger of which is 54-s and the smaller of which is 41-s (N. W. Gillham et al. in preparation). The mutant ac-20 has a great reduction in the proportion of chloroplast ribosome monomers (66-s + 70-s) and a modest increase in both chloroplast ribosome subunits compared to wild type. The mutant cr-1 shows a similarly great reduction in the proportion of chloroplast ribosome monomers and accumulates the large 54-s chloroplast ribosome subunit. Significantly, no small subunit peak is observed. The double mutant is quite similar in ribosome phenotype to ac-20 in that small 54-, 66-, and 70-s peaks are seen. However, no 41-s peak is observed suggesting that no free small subunits are present. Thus, the double mutant behaves like ac-20 in that it does not accumulate large amounts of the 54-s subunit whereas it behaves like cr-1 in that it does not accumulate detectable amounts of the 41-s subunit.

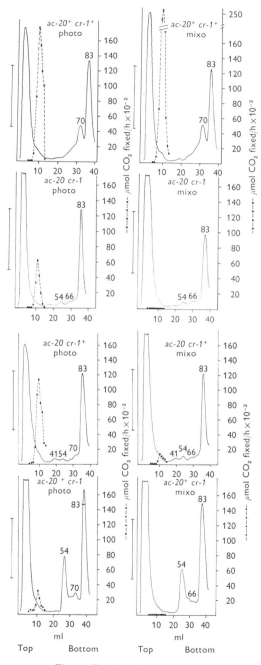

Fig. 2. For legend see opposite.

In earlier experiments with different isolates we did not detect a distinct 70-s peak for *ac-20*, *cr-1* and the double mutant and we did observe a very small 41-s peak for the double mutant (Boynton *et al.* 1970). However, the percentage of total chloroplast ribosome monomers in all 3 genotypes in the former study is in excellent agreement with that determined here (Table 3).

Numbers of ribosomes counted in electron micrographs of the chloroplast and cytoplasm of each of the 4 genotypes grown under mixotrophic and phototrophic conditions are summarized in Table 4, expressed both per μm^2 and per hypothetical cell. Large numbers of chloroplast ribosomes (*c.* 35% of total cell ribosomes) are formed by wild type cells grown either phototrophically or mixotrophically (Figs. 6, 8), whereas *ac-20* (Figs. 16, 19) and the double mutant (Figs. 11, 13) have fewer chloroplast ribosomes (less than 10% of total cell ribosomes) under both sets of growth conditions. In contrast, *cr-1* contains more chloroplast ribosome particles than either of the other mutant genotypes under both growth conditions (Figs. 22, 24). Most of the ribosome particles seen in *cr-1* are presumably not chloroplast ribosome monomers, but are the 54-s subunits accumulated by this mutant.

Accurate estimation of chloroplast ribosomes in electron micrographs of *cr-1* is more difficult than in the other genotypes since most of the ribosome particles in the chloroplast appear to be smaller, making them hard to distinguish from the stroma background. Mixotrophically grown cells of *cr-1* would be expected to have about the same number of chloroplast ribosome particles as phototrophically grown cells based on our sucrose gradient estimates (Table 3), and the 16% value estimated from electron micrographs is probably unrealistically low (Table 4).

Taking both the sucrose gradient and electron-microscopic ribosome estimates into account, the chloroplast ribosome monomers of both phototrophically and mixotrophically grown cells of *ac-20* and the double mutant constitute less than one fourth of the monomers observed in wild type. In *cr-1* the chloroplast ribosome monomers are about one third of the wild-type amount.

None of the mutant genotypes differ markedly from wild type in numbers of cytoplasmic ribosomes when grown either phototrophically or mixotrophically (Table 4). The somewhat reduced numbers of cytoplasmic ribosomes calculated per hypothetical cell in phototrophically grown *ac-20* result from relatively small cell size combined with unusually large nuclei in the 10 cells analysed yield a low estimate of net cytoplasmic area (Table 6). The number of ribosomes counted per unit area of cytoplasm was the same as in wild type (Table 4).

Quantitative estimates of the proportion of chloroplast and cytoplasmic ribosomes

---

Fig. 2. Sucrose gradient profiles of chloroplast and cytoplasmic ribosomes and the enzyme RuDPCase (partially purified) from phototrophically (photo) and mixotrophically (mixo) grown cells of the 4 genotypes of the tetratype tetrad 5·6 from the cross: *ac-20$^+$cr-1$^+$mt$^+$ × ac-20cr-1mt$^-$*. ———, absorbance at 254 nm (bar = 1·0 O.D. unit); – – –, μmol CO$_2$ fixed per h × 10^{-2}. In the 66-s and 70-s regions of the gradient only the higher of the 2 chloroplast ribosome monomer peaks is labelled. Extracts from 1·5 × 10^8 cells were loaded on all gradients except for those of the genotype *ac-20$^+$ cr-1* where 2·25 × 10^8 cells were loaded.

Table 3. *Relative amounts of ribosomes of the 4 genotypes of the 5·6 tetratype tetrad derived from the cross: ac-20⁺cr-1⁺mt⁺ × ac-20cr-1mt⁻. Mean values are given for cells grown under both phototrophic (P) and mixotrophic (M) conditions and the sedimentation velocities of the different species of ribosomes were determined by sucrose density-gradient centrifugation*

Genotype	Growth conditions	Sedimentation velocities and relative amounts of ribosomes as % of the total							No. of independent determinations
		Generic classes of ribosomes					Total chloroplast ribosome monomers (66 s + 70 s)	Total chloroplast ribosome monomers + subunits	
		83 s	70 s	66 s	54 s	41 s			
ac-20⁺ cr-1⁺	P	65	25	8	1	1	33	35	3
	M	61	29	7	2	1	36	39	3
ac-20 cr-1	P	89	5	3	3	0	8	11	2
	M	86	3	5	6	0	8	14	1
ac-20 cr-1⁺	P	87	4	2	3	4	6	13	3
	M	80	6	4	4	6	10	20	2
ac-20⁺ cr-1	P	60	11	3	26	0	14	40	3
	M	65	5	7	23	0	12	35	3

Table 4. *Numbers of chloroplast and cytoplasmic ribosomes estimated from electron micrographs of cells of the 4 genotypes from the tetratype tetrad 5·6 resulting from the cross: ac-20⁺cr-1⁺mt⁺ × ac-20cr-1mt⁻. Means with their standard errors are given for data from 10 cells of each genotype grown phototrophically (P) and mixotrophically (M)*

Genotype	Growth conditions	Ribosomes counted per μm^2 unobstructed section area			Total ribosomes estimated per cell section $\times 10^3$			Total ribosomes calculated per hypothetical cell $\times 10^5$		
		Cytoplasm	Chloroplast	% chloroplast	Cytoplasm	Chloroplast	% chloroplast	Cytoplasm	Chloroplast	% chloroplast
ac-20⁺ cr-1⁺	P	399 ± 21	302 ± 19	43 ± 2	8·01 ± 0·67	2·84 ± 0·43	26 ± 3	5·52 ± 0·68	3·53 ± 0·66	39 ± 0·4
	M	406 ± 13	245 ± 13	38 ± 1	10·2 ± 0·73	3·02 ± 0·28	23 ± 2	7·84 ± 0·85	3·80 ± 0·41	33 ± 2
ac-20 cr-1	P	430 ± 23	72 ± 4	14 ± 1	11·4 ± 0·90	0·82 ± 0·09	7 ± 1	9·31 ± 1·10	0·57 ± 0·09	6 ± 1
	M	381 ± 16	68 ± 11	15 ± 1	10·9 ± 0·69	0·97 ± 0·17	8 ± 1	8·71 ± 0·82	0·74 ± 0·20	7 ± 1
ac-20 cr-1⁺	P	394 ± 13	79 ± 8	16 ± 1	6·0 ± 0·46	0·61 ± 0·09	9 ± 1	3·60 ± 0·41	0·40 ± 0·08	10 ± 1
	M	339 ± 23	53 ± 6	14 ± 1	9·8 ± 0·75	0·73 ± 0·07	7 ± 1	7·45 ± 0·84	0·47 ± 0·08	7 ± 1
ac-20⁺ cr-1	P	415 ± 9	245 ± 24	36 ± 2	10·5 ± 0·67	3·40 ± 0·74	23 ± 3	8·18 ± 0·76	4·54 ± 1·36	32 ± 5
	M	371 ± 18	131 ± 13	26 ± 1	10·5 ± 0·93	1·69 ± 0·29	14 ± 1	8·35 ± 1·10	1·70 ± 0·41	16 ± 2

Table 5. Cytological organization and chlorophyll content of the 4 genotypes from the tetratype tetrad 5·6 resulting from the cross: ac-20^+cr-$1^+mt^+ \times ac$-$20\,cr$-$1\,mt^-$. Means with their standard errors are given for 10 cells of each genotype grown phototrophically (P) and mixotrophically (M)

| Genotype | Growth conditions | Area, μm²/median section | | | | | | % of cell area occupied by | | % of chloroplast area occupied by lamellar system | Total chlorophyll, μg/10⁶ cells |
		Cell	Net cytoplasm	Chloroplast	Net stroma	Chloroplast lamellar system	Mitochondria	chloroplast	mitochondria		
ac-20^+cr-1^+	P	46·9 ± 2·3	20·1 ± 1·3	17·5 ± 1·5	9·2 ± 1·0	6·9 ± 0·6	1·05 ± 0·13	37 ± 2	2·3 ± 0·3	40 ± 3	2·3
	M	61·1 ± 2·8	25·2 ± 1·9	21·5 ± 1·3	12·6 ± 1·2	8·0 ± 0·4	2·54 ± 0·17	35 ± 2	4·2 ± 0·3	38 ± 2	2·8
ac-$20\,cr$-1	P	60·8 ± 4·6	27·0 ± 2·3	20·0 ± 1·5	11·3 ± 1·1	7·9 ± 0·6	1·84 ± 0·35	33 ± 1	2·9 ± 0·5	40 ± 2	2·3
	M	73·9 ± 6·0	28·9 ± 1·6	27·0 ± 3·1	15·2 ± 2·4	9·6 ± 1·2	2·19 ± 0·31	36 ± 2	2·9 ± 0·3	36 ± 2	3·5
ac-$20\,cr$-1^+	P	49·2 ± 1·8	15·4 ± 1·2	20·4 ± 1·1	7·4 ± 0·4	11·4 ± 0·8	2·72 ± 0·69	42 ± 2	5·5 ± 1·3	56 ± 2	2·1
	M	71·8 ± 4·1	29·4 ± 2·2	23·4 ± 1·3	14·0 ± 0·8	8·4 ± 0·8	2·65 ± 0·30	33 ± 1	3·7 ± 0·4	35 ± 2	2·5
ac-20^+cr-1	P	60·4 ± 3·7	25·3 ± 1·7	22·7 ± 2·3	13·3 ± 1·7	8·9 ± 0·7	1·26 ± 0·26	37 ± 3	2·2 ± 0·5	40 ± 2	2·9
	M	67·6 ± 6·5	28·8 ± 2·7	23·7 ± 2·9	13·2 ± 1·7	8·0 ± 1·1	1·80 ± 0·01	35 ± 3	0·6 ± 0·001	34 ± 2	2·7

Table 6. Activities of crude extracts and partially purified ribulose diphosphate carboxylase isolated from cells of the 4 genotypes from the tetratype tetrad 5·6 resulting from the cross: ac-20^+cr-$1^+mt^+ \times ac$-$20\,cr$-$1\,mt^-$. Mean values are given for cells of each genotype grown phototrophically (P) and mixotrophically (M). Values given in parentheses may not be significantly above background

| Genotype | Growth conditions | Crude extracts | | | | | Partially purified enzyme | | | | |
| | | Activity/10⁹ cells | | Specific activity | | No. of independent experiments | Activity/10⁹ cells | | Specific activity | | No. of independent determinations |
		μmol CO₂ fixed/h	% wild type	μmol CO₂ fixed/h mg protein	% wild type		μmol CO₂ fixed/h	% wild type	μmol CO₂ fixed/h mg protein	% wild type	
ac-20^+cr-1^+	P	62	100	5·0	100	3	32	100	16	100	2
	M	69	100	5·5	100	1	42	100	15	100	2
ac-$20\,cr$-1	P	13	21	1·6	32	3	11	34	5	31	2
	M	(0·5)	(0·007)	(0·04)	(0·07)	2	0	0	0	0	2
ac-$20\,cr$-1^+	P	45	73	4·6	92	3	41	128	13	81	1
	M	8	12	0·5	9	1	3	7	1	7	1
ac-20^+cr-1	P	6	10	1·1	22	4	4	13	3	19	2
	M	(0·4)	(0·006)	(0·06)	(0·01)	2	0	0	0	0	2

determined by electron microscopy (i.e. per hypothetical cell, Table 4) and sucrose-gradient sedimentation (i.e. total chloroplast ribosome monomers + subunits, Table 3) are in reasonably good agreement in all cases. Precise agreement between the methods cannot be expected for reasons we have enumerated in an earlier paper (Bourque *et al.* 1971).

Amounts and activities of RuDPCase of the 4 genotypes

A large peak of activity for RuDPCase is seen in the 18- to 21-s region of sucrose gradients designed to separate the various ribosome species of the wild-type cell (Figs. 1, 2). The total enzyme activity of mixotrophically grown cells of wild type is slightly higher than that observed in cells grown phototrophically. Neither *cr-1* nor the double mutant exhibit a detectable activity peak when grown mixotrophically whereas *ac-20* shows a very small activity peak (about 7 % of wild type) under these growth conditions (Table 6). All 3 mutant genotypes have a clearly detectable RuDPCase activity peak when grown phototrophically (Fig. 2). In the single determination made for *ac-20* the activity peak did not differ markedly from the mean value obtained for wild type (Table 6). In *cr-1* the activity peak is 13 % of wild type and for the double mutant it is 34 % of wild type (Table 6). The RuDPCase activity peak in all mutant genotypes appears in the same position in the sucrose gradient as that of wild type.

The foregoing observations raise 2 questions. First, do the 3 mutant genotypes actually fail to make the enzyme RuDPCase when grown mixotrophically or do they synthesize an inactive enzyme in wild-type amounts? Secondly, does the RuDPCase made by phototrophically grown cells of the 3 mutant genotypes have a lower activity per unit of enzyme protein than the enzyme made by wild type? We sought to answer these questions by establishing the amount of RuDPCase made by the 4 genotypes and comparing the enzyme's specific activity at different levels of purification.

The relative amounts of RuDPCase made were measured by highly purifying the enzyme on sucrose gradients so that it separated as a distinct peak from the material at the top of the gradients (Fig. 3) and its absorbance at 254 nm was then determined (Table 7). Wild-type cells grown either phototrophically or mixotrophically have a distinct enzyme peak which constitutes 8–10 % of the total soluble cell protein. Cells of *ac-20* also have a large enzyme peak (43 % of wild type) when grown photo-trophically, but a smaller peak (28 % of wild type) when grown mixotrophically. The *cr-1* mutant has no observable optical density peak when grown mixotrophically but has a small peak which cannot be measured accurately in the expected position when grown phototrophically. The double mutant shows a small RuDPCase peak (10 % of

---

Fig. 3. Sucrose gradient profiles of highly purified RuDPCase from phototrophically (photo) and mixotrophically (mixo) grown cells of the 4 genotypes of the tetratype tetrad 5·6 from the cross: ac-20^+cr-$1^+mt^+ \times ac$-$20cr$-$1mt^-$. ——; absorbance at 254 nm (bar = 0·1 O.D. unit). The RuDPCase peak, when present appears in about the 20-ml region on all gradients. Extracts from 8×10^8 cells were loaded on all gradients except for that of the genotype ac-20^+cr-1 grown phototrophically where 10×10^8 cells were loaded.

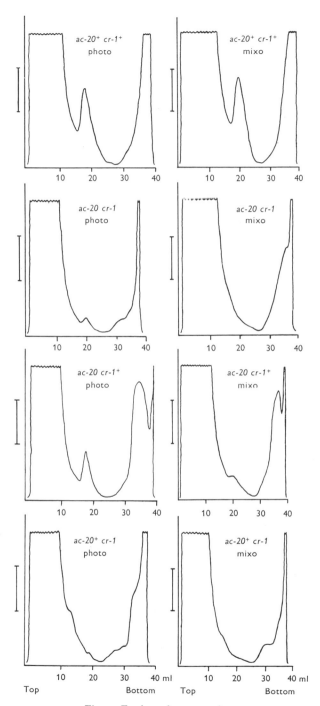

Fig. 3. For legend see opposite.

wild type) when grown phototrophically, but none when grown mixotrophically. Clearly, the 3 mutant genotypes make far less enzyme than wild type or no enzyme at all when grown mixotrophically whereas each can make at least some enzyme when grown phototrophically.

The specific activity of purified RuDPCase can be measured in phototrophically grown cells of all 4 genotypes (Table 8). For mixotrophically grown cells, meaningful specific activities can be obtained only for wild type and *ac-20* since in *cr-1* and the double mutant little if any RuDPCase can be detected (Table 8, Fig. 3). We have compared specific activities and total activities/10^9 cells in crude extracts of cells

Table 7. *Amounts of ribulose diphosphate carboxylase isolated from cells of the 4 genotypes from the tetratype tetrad 5·6 resulting from the cross: ac-20⁺ cr-1⁺ mt⁺ × ac-20 cr-1 mt⁻. Mean values are given for cells of each genotype grown phototrophically (P) and mixotrophically (M) in the experiments summarized in Tables 6 and 8*

		mg protein/10^9 cells				
Genotype	Growth conditions	Crude homogenate	% wild type	Purified enzyme	% wild type	Enzyme as % of cell protein
ac-20⁺ cr-1⁺	P	12·2	100	1·2	100	10
	M	12·5	100	1·0	100	8
ac-20 cr-1	P	8·3	68	0·12	10	1
	M	11·8	94	*	—	—
ac-20 cr-1⁺	P	10·8	89	0·52	43	4
	M	16·3	130	0·28	28	2
ac-20⁺ cr-1	P	8·2	67	*	—	—
	M	7·9	63	*	—	—

* No measurable peak resolvable on gradient.

(Table 6), in partially purified enzyme preparations (Table 6) and in highly purified RuDPCase (Table 8). The enzyme loses between 50 and 75 % of its activity during purification by our method (Table 8). Assuming no loss in activity we would expect about a 10-fold increase in specific activity upon complete purification, since RuDPCase already constitutes about 10 % of the total soluble protein in crude extracts of wild type (Table 7). In the mutants we would expect to see a somewhat greater increase in specific activity upon complete purification since RuDPCase constitutes a smaller percentage of the total soluble protein (Table 7).

Purification of RuDPCase from both phototrophically and mixotrophically grown cells of wild type resulted in a 4- to 5-fold increase in specific activity of the enzyme to a final value of about 23 μmol CO_2 fixed/h/mg protein (Table 8). Michaelis constants of $3\cdot0 \times 10^{-4}$ for the RuDP and $1\cdot5 \times 10^{-2}$ for the HCO_3^- substrates determined for the highly purified enzyme from phototrophically grown cells of wild type (Table 9) agree well with values published for higher plants by Kawashima & Wildman (1970).

RuDPCase from *ac-20* grown phototrophically was purified to about the same degree as that from wild type (Table 8) and had a similar specific activity and Michaelis

Table 8. *Activities of highly purified ribulose diphosphate carboxylase isolated from cells of the 4 genotypes from the tetratype tetrad 5·6 resulting from the cross: ac-20^+ cr-1^+ mt^+ × ac-20 cr-1 mt^-. Mean values are given for cells of each genotype grown phototrophically (P) and mixotrophically (M). Values given in parentheses may not be significantly above background*

| Genotype | Growth conditions | Activity/10^9 cells | | Specific activity | | Increase in specific activity during purification | Predicted specific activity*, μmol CO_2/h/ mg protein | Estimated loss of activity during puri- fication† | No. of independent experiments |
		μmol CO_2 fixed/h/ peak tube	% wild type	μmol CO_2 fixed/h/ mg protein	% wild type				
ac-20^+ cr-1^+	P	5·4	100	24	100	4·8	52	54	3
	M	4·5	100	22	100	4·0	69	68	2
ac-20 cr-1	P	0·97	18	16	67	10·0	—	—	3
	M	(0·06)	(1)	(0·6)	(3)	15·0	—	—	1
ac-20 cr-1^+	P	2·3	43	21	87	4·6	86	76	4
	M	0·9	20	8	36	16·0	29	72	1
ac-20^+ cr-1	P	0·28	5	2	8	1·8	—	—	2
	M	(0·03)	(0·6)	(0·7)	(3)	11·7	—	—	1

* $\dfrac{\mu\text{mol } CO_2 \text{ fixed/h/}10^9 \text{ cells (crude extract)}}{\text{mg purified enzyme/}10^9 \text{ cells}}$. This calculation assumes no loss of enzyme protein during purification by centrifugation.

† $\dfrac{\text{Predicted specific activity–specific activity measured on highly purified enzyme}}{\text{predicted specific activity}} \times 100$.

constants for RuDP and HCO_3^- (Table 9). Thus RuDPCase made by the *ac-20* mutant is functionally the same as that of wild type, but is made in reduced amounts.

While the specific activity of RuDPCase from mixotrophically grown cells of *ac-20* increases 5-fold upon purification, it fails to reach the wild type value (Table 8). Whether this reflects more residual protein contamination in our preparation from *ac-20* (Fig. 3) or an unusually high loss in activity in the single experiment conducted is not known. The reduction in protein in our highly purified enzyme preparation (28 % of wild type) parallels the reduction in CO_2-fixing ability of the enzyme in the crude extracts (12 % of wild type). Clearly, mixotrophically grown *ac-20* cells make less RuDPCase than phototrophically grown cells.

Table 9. *Michaelis constants (K_m) for ribulose 1,5-diphosphate and HCO_3^- of highly purified ribulose diphosphate carboxylase isolated from cells of the 4 genotypes of the tetratype tetrad 5·6 resulting from the cross: ac-20+ cr-1+ mt+ × ac-20 cr-1 mt-. Mean values are given for cells grown phototrophically*

Genotype	K_m (RuDP)	No. of independent determinations	K_m (HCO_3^-)	No. of independent determinations
ac-20+ cr-1+	$3·0 \times 10^{-4}$	2	$1·5 \times 10^{-2}$	2
ac-20 cr-1	$1·5 \times 10^{-4}$	1	$2·8 \times 10^{-2}$	1
ac-20 cr-1+	$2·3 \times 10^{-4}$	3	$2·0 \times 10^{-2}$	1
ac-20+ cr-1	*	2	$2·0 \times 10^{-2}$	1

* Amounts of enzyme obtained were so small that reasonable K_m (RuDP) could not be measured.

Attempts to purify RuDPCase from cells of *cr-1* or the double mutant grown mixotrophically have met with no success because the *cr-1* mutation, by itself or in combination with *ac-20* does not make detectable amounts of this enzyme (Table 7, Fig. 3). The exceedingly low CO_2-fixing ability of the enzyme measured in crude extracts and in the highly purified preparations from these mixotrophically grown cells is probably not significantly above background (Tables 6, 8), since fraction by fraction analysis of partially purified RuDPCase from either genotype yielded no distinct activity peak (Fig. 2).

While *cr-1* and the double mutant do not synthesize RuDPCase under mixotrophic growth conditions, they make modest amounts of the enzyme under phototrophic conditions. Attempts at purification of the RuDPCase from *cr-1* yielded about a 2-fold increase in specific activity over that observed in the crude extracts (Tables 6, 8). The exceedingly small RuDPCase peak in highly purified preparations is so heavily contaminated by proteins from the top of the gradient (Fig. 3) that comparisons of the specific activity to wild type are meaningless. The enzyme does appear to be functionally similar to that from wild type in terms of the Michaelis constant for the HCO_3^- substrate but we have been unable to determine the Michaelis constant for RuDP accurately.

Purification of RuDPCase from cells of the double mutant grown phototrophically was more successful, yielding a specific activity that approaches the wild-type value (Fig. 3, Table 8). On the basis of similarities in specific activity and in Michaelis constants for the RuDP and HCO_3^- substrates (Table 9), we believe that the RuDPCase formed by the double mutant is functionally similar to the wild-type enzyme. These results suggest that neither the *cr-1* nor the *ac-20* mutations alter the functional properties of the RuDPCase enzyme significantly, but instead only influence the amount synthesized.

Chloroplast ribosome phenotype and the structural organization of the chloroplast

The 3 mutant genotypes grown under both phototrophic and mixotrophic conditions form a chloroplast which equals or exceeds in size that formed by wild type (Table 5; Figs. 4, 7, 9, 12, 14, 17, 20, 23). Both the extent of the lamellar system and the amount of chlorophyll in the chloroplasts of all mutant genotypes are similar to wild type (Table 5). Therefore the reduction in chloroplast ribosome monomers in the mutants *ac-20* and *cr-1* and in the double mutant appears to have no direct effect on the quantitative formation of these major chloroplast components.

However, the organization of the lamellar system is profoundly altered in all 3 mutant genotypes in the same fashion when they are grown under mixotrophic conditions (Figs. 12, 17, 23). In each case the lamellae are either stacked into giant grana several microns in diameter (Fig. 18) and up to 20 disks high or they exist in a single unpaired state. Pyrenoids are conspicuously absent and the chloroplast stroma has a lower density than that of wild type. The so-called 'clear' areas in the stroma commonly reported to contain DNA appear to be far more extensive than in wild type. Our observation that *ac-20* grown mixotrophically forms conspicuous giant grana is at variance with the report of Goodenough & Levine (1970) that the same mutant genotype forms only unpaired lamellae under apparently similar growth conditions.

When all 3 mutant genotypes are grown phototrophically, their chloroplast lamellar systems approach wild type in organization (Figs. 4, 9, 14, 20). Two-disk grana predominate over much of the chloroplast, but in localized areas the grana stacks may consist of up to 5 disks. Pyrenoids are also formed (Figs. 4, 9, 15, 21), and the DNA-containing areas of the stroma appear to be smaller than those of mixotrophically grown cells. The chloroplast phenotypes of all 3 mutant genotypes grown phototrophically are the same as described earlier for our *ac-20 mt⁻* stock (Bourque *et al.* 1971) and by Goodenough & Levine (1970) for *ac-20*.

Differences in mitochondrial conformation in mixotrophically and phototrophically grown cells of Chlamydomonas

Cells of wild type and the 3 mutant genotypes grown phototrophically form one or more large mitochondrial aggregates in contrast to many small mitochondria observed in mixotrophically grown cells (Figs. 5, 10, 14, 20). The total mitochondrial area measured per cell section does not change significantly (Table 6) but the number of mitochondria per cell drops markedly. The reason for these differences in mitochondrial conformation under phototrophic growth conditions is presently unknown.

Two approaches are useful in trying to understand whether chloroplast ribosomes promote the translation of unique sets of messengers. The first employs inhibitors specific for translation on chloroplast ribosomes and the second, mutants having a chloroplast ribosome deficiency or mutants defective in chloroplast ribosome function. Ideally, both approaches should give concordant results. Components synthesized in the presence of the inhibitor also should be synthesized in normal amounts by these mutants whereas components whose synthesis is blocked by the inhibitor should be synthesized in reduced amounts if at all by the mutants. Furthermore, all inhibitors affecting chloroplast ribosome function should give qualitatively similar results and so should all chloroplast ribosome mutants.

Levine and his co-workers (Armstrong, Surzycki, Moll & Levine, 1971; Goodenough & Levine, 1970; Levine & Paszewski, 1970; Surzycki, Goodenough, Levine & Armstrong, 1970; Togasaki & Levine, 1970) have used a combination of these approaches in *C. reinhardi*. Armstrong *et al.* (1971) examined the effects of the chloroplast ribosome inhibitors chloramphenicol and spectinomycin as well as the inhibitor of cytoplasmic ribosome function cycloheximide on the formation of various chloroplast components in wild-type cells grown synchronously. Under these conditions the amounts or activities of each component studied increase 2- to 3-fold during the same period in the cycle. Two of the three possible patterns of inhibition have been observed by Armstrong *et al.* (1971). First, increases in chlorophyll, ferridoxin, ferridoxin-NADP reductase, and phosphoribulokinase were blocked only by cycloheximide. The authors concluded that the synthesis of these components required translation of messenger RNA only on cytoplasmic ribosomes. Secondly, increases in RuDPCase, cytochrome 563 and probably cytochrome 553 were inhibited by cycloheximide, chloramphenicol and spectinomycin, suggesting that all 3 components require translational activity of both chloroplast and cytoplasmic ribosomes. Significantly, Armstrong and co-workers did not find a single chloroplast component whose synthesis was blocked only by chloramphenicol and spectinomycin, but not by cycloheximide.

The mutant approach to chloroplast ribosome function in *C. reinhardi* has been confined to studies of the chloroplast ribosome deficient mutant *ac-20* (Goodenough & Levine, 1970; Levine & Paszewski, 1970; Togasaki & Levine, 1970). Those components whose synthesis is predicted to occur on cytoplasmic ribosomes from the inhibitor studies with wild type are present in normal or near normal amounts in mixotrophically grown *ac-20* whereas those components whose synthesis seems to require functioning of both chloroplast and cytoplasmic ribosomes are present in reduced amounts. It should be noted that Armstrong *et al.* (1971) have now found the expected reduction in amounts of cytochrome 553 and 563 in *ac-20* grown mixotrophically but they had not reported this reduction earlier (Goodenough & Levine, 1970; Levine & Paszewski, 1970). Thus for mixotrophically grown cells of *ac-20* results of the 2 experimental approaches are concordant.

Since the mutant genotypes *ac-20*, *cr-1* and *ac-20cr-1* all cause a reduction in the number of chloroplast ribosome monomers per cell and quite possibly also a reduction in ribosome activity, we would predict that they would have qualitatively although not necessarily quantitatively similar effects on chloroplast phenotype. The enzyme RuDPCase served as a marker for chloroplast ribosome function. Inhibitor studies in both *Chlamydomonas* (Armstrong *et al.* 1971) and barley (Criddle, Dau, Kleinkopf & Huffaker, 1970) suggest that RuDPCase synthesis requires the active participation of both chloroplast and cytoplasmic ribosomes. This is reasonable since RuDPCase is a protein of high molecular weight (*c.* 500000) composed of an aggregate of 2 types of subunits having molecular weights of 10000 and 20000 respectively (Akazawa, 1970; Kawashima & Wildman, 1970). In fact Criddle *et al.* (1970) have claimed that chloramphenicol specifically inhibits synthesis of the large subunit of RuDPCase in barley whereas the synthesis of the small subunit is inhibited by cycloheximide.

Precedents for ribosomal specificity in the translation of messenger RNA now exist. For example, Lodish & Robertson (1969) have shown that when RNA derived from the RNA-containing bacteriophage f2 is used as a synthetic messenger, all 3 initiator sites for translation are recognized by ribosomes from the bacterium *Escherichia coli*, but only one of these is recognized by the ribosomes of *Bacillus stearothermophilus*, and none by the ribosomes of *Bacillus subtilis*.

All 3 mutant genotypes of *C. reinhardi* deficient in chloroplast ribosome monomers make either reduced amounts or no RuDPCase when grown mixotrophically – in excellent concordance with results of the inhibitor studies. We might also predict that both the inhibitor and mutant approaches would effect qualitatively similar alterations in chloroplast organization if specific proteins necessary for the ordering of the lamellar system into wild-type grana were translated on chloroplast ribosomes. We have partially verified these predictions, since all 3 chloroplast ribosome deficient genotypes grown mixotrophically have similar abnormally organized chloroplasts characterized by giant grana. Our observation that chlorophyll content in mixotrophically grown cells of the 3 mutant genotypes is the same as in wild type agrees well with the findings of the inhibitor studies (Armstrong *et al.* 1971) and *ac-20* (Goodenough & Levine, 1970) which suggest that chlorophyll synthesis is mediated solely by cytoplasmic ribosomes.

All 3 mutant genotypes grown phototrophically approach wild type in terms of RuDPCase and chloroplast ultrastructure, but their chloroplast ribosomes do not increase significantly. They exhibit RuDPCase activity although marked quantitative differences are observed in the amounts of enzyme present. Whereas *ac-20* makes half the enzyme protein detected in wild type, the amount made by *cr-1* is measurable only as an activity peak and the amount made by double mutant is about 10 % of wild type (Tables 7, 8). Chloroplast organization also approaches wild type in that a pyrenoid is formed and the lamellae are paired into 2- to 3-disk grana.

Normalization of the chloroplast organization and level of RuDPCase in *ac-20* grown phototrophically has been explained by Goodenough & Levine (1970) in terms of the increase in chloroplast ribosomes which they observed upon transfer of their stock from mixotrophic to phototrophic growth conditions. Certainly, if more chloroplast ribosomes are available for translation of specific messengers, then an increase in chloroplast ribosomes could lead to an increase in the synthesis of the chloroplast components translated on those ribosomes. However, our observations have consistently shown that while there are no changes in numbers of chloroplast ribosomes in the 3 mutant genotypes between our conditions of mixotrophic and phototrophic growth, large changes in the level of RuDPCase and in chloroplast phenotype occur. Thus, in *ac-20*, any correlations between an increase in chloroplast ribosomes, an increase in RuDPCase and a normalization of chloroplast organization are fortuitous.

How can we preserve the argument that the pleiotropic effects of the *ac-20* and *cr-1* mutants grown mixotrophically result from a defect in chloroplast protein synthesis if an increase in chloroplast ribosome number does not occur upon transfer of the cells to phototrophic conditions ? There are 3 possibilities. First, when the cells are grown phototrophically the cytoplasmic ribosomes might pre-empt functions specific for chloroplast ribosomes under mixotrophic conditions. Secondly, the increase in doubling times observed in all 4 genotypes under phototrophic conditions might allow additional time for the synthesis of those components translated on chloroplast ribosomes. Thirdly, the rate at which chloroplast ribosomes translate certain messages might change when cells are shifted from mixotrophic to phototrophic growth conditions.

The possibility that cytoplasmic ribosomes are responsible for the synthesis of intact molecules of RuDPCase in the mutant genotypes grown phototrophically appears unlikely, since each has a different chloroplast ribosome phenotype and synthesizes different amounts of RuDPCase, but all appear to have normal amounts of cytoplasmic ribosomes with normal sedimentation velocity. If the cytoplasmic ribosomes were responsible, their ability to function would have to differ between genotypes. This would mean that both cytoplasmic and chloroplast ribosomes were altered simultaneously in all 3 mutants, which seems highly improbable genetically. The inhibitor experiments of Armstrong *et al.* (1971) carried out on wild-type cells grown synchronously in the absence of acetate on an alternating light-dark cycle show that the synthesis of RuDPCase is inhibited by chloramphenicol and spectinomycin under these, essentially, phototrophic growth conditions. This suggests that synthesis of RuDPCase requires functioning chloroplast ribosomes under phototrophic conditions. Furthermore, Togasaki (cited in Armstrong *et al.* 1971) has found that the synthesis of RuDPCase in phototrophically grown cells of *ac-20* can be inhibited by chloramphenicol. If the mutant synthesized the enzyme solely on cytoplasmic ribosomes this should not occur.

The mutant genotypes might also approach wild type when grown phototrophically because of their increased doubling times under these conditions. We detect doubling time increases of less than 1·5-fold under phototrophic conditions for all 3 genotypes (P/M, Table 2) whereas both the activity and amount of RuDPCase measured increase at least several-fold (Tables 6–8). Therefore the increase in the length of the doubling time under phototrophic conditions is not sufficient to account for the observed increase in activity and amount of RuDPCase in the 3 mutant genotypes since any increase in the component greater than a doubling must be attributed to causes other than a simple doubling of the time between divisions.

We propose that cells of the 3 mutant genotypes approach wild type under phototrophic conditions due to an increase in the rate at which their chloroplast ribosomes translate specific messengers. Such a rate change could occur either because the chloroplast ribosomes become programmed to translate preferentially a certain set of messengers under phototrophic conditions or because transcription of these messengers under phototrophic conditions is preferentially increased so that these messengers form a larger fraction of the total population in the chloroplast under phototrophic conditions and compete more effectively for the existing chloroplast ribosomes. We believe we do not detect a similar increase in RuDPCase in phototrophically grown cells of wild type as compared to mixotrophically grown cells because the amount of RuDPCase already present in mixotrophically grown cells is sufficiently high for maximal CO_2 fixation under the light intensities used in our experiments. Light intensity may be limiting photosynthetic rate in our wild-type cultures and if we were to increase the light intensity we might expect to detect higher levels of RuDPCase in phototrophically than in mixotrophically grown cells of wild type. Using somewhat higher light intensities Togasaki & Levine (1970) report that their phototrophically grown wild-type cultures consistently had somewhat higher activities for RuDPCase than the mixotrophic cultures.

The foregoing hypothesis predicts that the chloroplast ribosomes observed in ac-20, cr-1 and the double mutant are able to mediate protein synthesis. Why then do we observe quantitative differences between the 3 mutant genotypes in terms of their ability to synthesize RuDPCase under phototrophic conditions ? We can account for these differences simply by assuming that ac-20 and cr-1 affect chloroplast ribosomes differently (see Boynton et al. 1970). Chloroplast ribosome function in cr-1 appears to be more drastically impaired than in ac-20.

A wide variation in RuDPCase (Fraction I) protein content has also been observed in 3 albina and 21 xantha mutants of barley which are known to show Mendelian inheritance and to have a variety of defects in their chloroplast organization or development (von Wettstein, Henningsen, Boynton, Kannangara & Nielsen, 1971). However, the specific activity of the RuDPCase purified by Sephadex gel filtration was practically identical to that of wild type, indicating that none of the mutants affected the structural gene(s) for this enzyme. It is not yet known whether the mutants have altered or deficient chloroplast ribosomes and as a consequence of this form sub-normal amounts of the enzyme. Kleinhofs & Shumway (1969) have described 15 albina mutants of barley having 1 % or less of the wild-type level of RuDPCase activity and 8 chlorina mutants with varying amounts of chlorophyll and subnormal levels of RuDPCase activity. One of the chlorina mutants having the lowest RuDPCase activity is known to be inherited in a non-Mendelian (cytoplasmic) manner whereas the other mutants are presumed to be Mendelian in inheritance. These authors also mention that the chloroplast stroma in electron micrographs of the mutants with low RuDPCase activity has a characteristic low electron-density similar to that which we observe in the mutants of C. reinhardi deficient in chloroplast ribosomes and RuDPCase. It will be of great interest to learn whether any of these barley mutants are deficient in RuDPCase because of a defect in their chloroplast protein synthesizing system.

We would like to thank Miss Nancy Alexander for her skilful assistance in carrying out the genetic aspects of the present research. This work was supported in part by grants from the National Science Foundation, GU–2018 and GB–22769.

REFERENCES

AKAZAWA, T. (1970). The structure and function of fraction-I protein. Regulatory aspects of photosynthetic CO_2-fixation in chloroplasts. *Progress in Phytochemistry* **2**, 107–141.
ARMSTRONG, J. J., SURZYCKI, S. J., MOLL, B. & LEVINE, R. P. (1971). Genetic transcription and translation specifying chloroplast components in *Chlamydomonas reinhardi*. *Biochemistry, N.Y.* **10**, 692–701.
ARNON, D. I. (1949). Copper enzymes in isolated chloroplasts. Polyphenoloxidase in *Beta vulgaris*. *Pl. Physiol., Lancaster* **24**, 1–15.
BJÖRKMAN, O. (1968). Carboxydismutase activity in shade-adapted and sun-adapted species of higher plants. *Physiol. Pl.* **21**, 1–10.
BJÖRKMAN, O. & GAUHL, E. (1969). Carboxydismutase activity in plants with and without β-carboxylation photosynthesis. *Planta* **88**, 197–203.
BOURQUE, D. P., BOYNTON, J. E. & GILLHAM, N. W. (1971). Studies on the structure and cellular location of various ribosome and ribosomal RNA species in the green alga *Chlamydomonas reinhardi*. *J. Cell Sci.* **8**, 153–183.
BOYNTON, J. E., GILLHAM, N. W. & BURKHOLDER, B. (1970). Mutations altering chloroplast ribosome phenotype in *Chlamydomonas*. II. A new Mendelian mutation. *Proc. natn. Acad. Sci. U.S.A.* **67**, 1505–1512.
BRAY, G. A. (1960). A simple efficient liquid scintillator for counting aqueous solutions in a liquid scintillation counter. *Analyt. Biochem.* **1**, 279–285.
CRIDDLE, R. S., DAU, B., KLEINKOPF, G. E. & HUFFAKER, R. C. (1970). Differential synthesis of ribulose diphosphate carboxylase subunits. *Biochem. biophys. Res. Commun.* **41**, 621–627.
EBERSOLD, W. T. & LEVINE, R. P. (1959). A genetic analysis of linkage group I of *Chlamydomonas reinhardi*. *Z. VererbLehre* **90**, 74–82.
GILLHAM, N. W. (1965). Induction of chromosomal and non-chromosomal mutations in *Chlamydomonas reinhardi* with N-methyl-N′-nitro-N-nitrosoguanidine. *Genetics* **52**, 529–537.
GOLDTHWAITE, J. J. & BOGORAD, L. (1971). A one-step method for the isolation and determination of leaf ribulose-1,5-diphosphate carboxylase. *Analyt. Biochem.* **41**, 57–66.
GOODENOUGH, U. W. & LEVINE, R. P. (1970). Chloroplast structure and function in *ac-20*, a mutant strain of *Chlamydomonas reinhardi*. III. Chloroplast ribosomes and membrane organization. *J. Cell Biol.* **44**, 547–562.
HASTINGS, P. J., LEVINE, E. E., COSBEY, E., HUDOCK, M. O., GILLHAM, N. W., SURZYCKI, S. J., LOPPES, R. & LEVINE, R. P. (1965). The linkage groups of *Chlamydomonas reinhardi*. *Microbial Genet. Bull.* **23**, 17–19.
HOOBER, J. K. & BLOBEL, G. (1969). Characterization of the chloroplast and cytoplasmic ribosomes of *Chlamydomonas reinhardi*. *J. molec. Biol.* **41**, 121–138.
KAWASHIMA, N. & WILDMAN, S. G. (1970). Fraction I protein. *A. Rev. Pl. Physiol.* **21**, 325–358.
KLEINHOFS, A. & SHUMWAY, L. K. (1969). Correlation of ribulose 1,5-diphosphate carboxylase activity with chlorophyll content and ultrastructure in induced mutants of *Hordeum vulgare*. *Biochem. Genet.* **3**, 485–492.
LEVINE, R. P. & GOODENOUGH, U. W. (1970). The genetics of photosynthesis and the chloroplast in *Chlamydomonas reinhardi*. *A. Rev. Genet.* **4**, 397–408.
LEVINE, R. P. & PASZEWSKI, A. (1970). Chloroplast structure and function in *ac-20*, a mutant strain of *Chlamydomonas reinhardi*. II. Photosynthetic electron transport. *J. Cell Biol.* **44**, 540–546.

LEVINE, R. P. & TOGASAKI, R. K. (1965). A mutant strain of *Chlamydomonas reinhardi* lacking ribulose diphosphate carboxylase activity. *Proc. natn. Acad. Sci. U.S.A.* **53**, 987–990.

LINEWEAVER, H. & BURKE, D. (1934). The determination of enzyme dissociation constants. *J. Am. Chem. Soc.* **56**, 658–666.

LODISH, H. F. & ROBERTSON, H. D. (1969). Regulation of *in vitro* translation of bacteriophage f_2 RNA. *Cold Spring Harb. Symp. quant. Biol.* **34**, 655–673.

LOWRY, O. H., ROSEBROUGH, N. J., FARR, A. L. & RANDALL, R. J. (1951). Protein measurement with the Folin phenol reagent. *J. biol. Chem.* **193**, 265–275.

SUEOKA, N. (1960). Mitotic replication of deoxyribonucleic acid in *Chlamydomonas reinhardi*. *Proc. natn. Acad. Sci. U.S.A.* **46**, 83–91.

SURZYCKI, S. J., GOODENOUGH, U. W., LEVINE, R. P. & ARMSTRONG, J. J. (1970). Nuclear and chloroplast control of chloroplast structure and function in *Chlamydomonas reinhardi*. *Symp. Soc. exp. Biol.* **24**, 13–37.

TOGASAKI, R. K. & LEVINE, R. P. (1970). Chloroplast structure and function in *ac-20*, a mutant strain of *Chlamydomonas reinhardi*. I. CO_2 fixation and ribulose-1,5-diphosphate carboxylase synthesis. *J. Cell Biol.* **44**, 531–539.

VON WETTSTEIN, D., HENNINGSEN, K. W., BOYNTON, J. E., KANNANGARA, G. C. & NIELSEN, O. F. (1971). The genetic control of chloroplast development in barley. In *Autonomy and Biogenesis of Mitochondria and Chloroplasts* (ed. N. K. Boardman), pp. 205–223. Amsterdam: North-Holland Publishing Co.

ABBREVIATIONS ON PLATES

cr and *pr*, ribosomes in the cytoplasm and chloroplast, respectively; *pe*, double-membrane plastid envelope.

Fig. 4. Section through a typical *ac-20⁺cr-1⁺* (wild type) cell of *C. reinhardi* grown phototrophically. The cup-shaped chloroplast with a well-developed lamellar system and pyrenoid occupies the periphery of the cell. × 12600.

Fig. 5. Section through a large branched mitochondrial aggregate of a phototrophically grown wild-type cell. × 39000.

Fig. 6. A portion of the wild-type cell of Fig. 4 at high magnification showing ribosomes in both the cytoplasm and the chloroplast. × 150000.

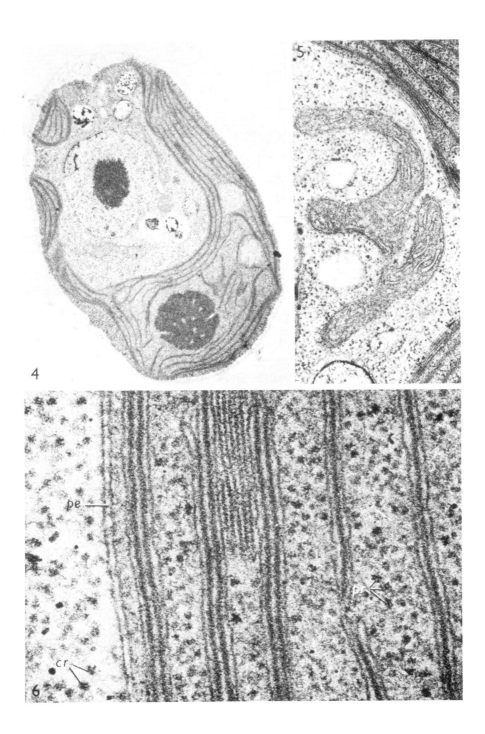

Fig. 7. Section through a typical *ac-20+cr-1+* (wild type) cell grown mixotrophically. Although a pyrenoid is not seen in this section through the chloroplast, it is present in wild-type cells grown under mixotrophic conditions. × 12600.

Fig. 8. A portion of the wild-type cell of Fig. 7 at high magnification showing ribosomes in both the cytoplasm and the chloroplast. × 150000.

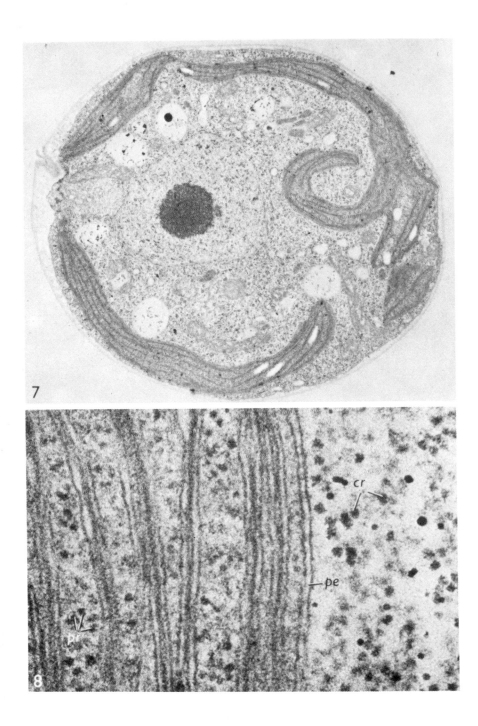

Fig. 9. Section through a typical cell of the double mutant *ac-20cr-1* grown photo-trophically. The lamellar system of the cup-shaped chloroplast is organized largely into 2-disk grana, and a well formed pyrenoid is evident. × 12600.

Fig. 10. Section through a mitochondrial aggregate of a phototrophically grown cell of the double mutant. × 21000.

Fig. 11. A portion of the double mutant cell of Fig. 9 at high magnification showing ribosomes in the cytoplasm and their relative absence in the chloroplast. × 150000.

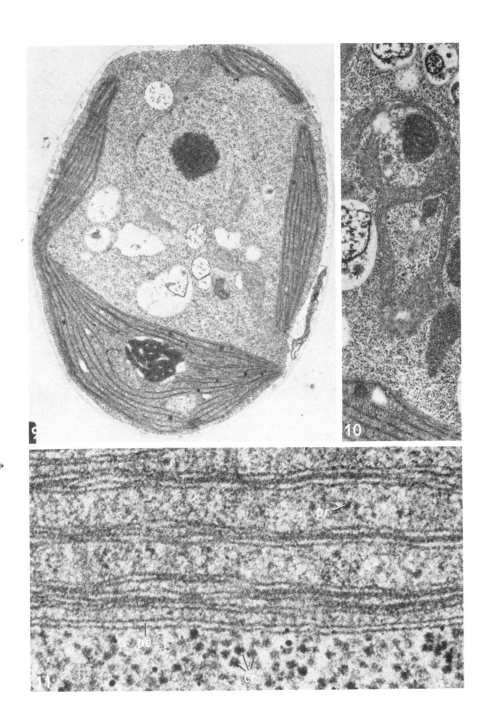

Fig. 12. Section through a typical cell of the double mutant *ac-20cr-1* grown mixo-trophically. The extensive lamellar system in the chloroplast exists either as single, unpaired disks or as giant grana. No pyrenoid is formed. × 12600.

Fig. 13. A portion of the double mutant cell of Fig. 12 at high magnification showing ribosomes in the cytoplasm and their relative absence in the chloroplast. × 150000.

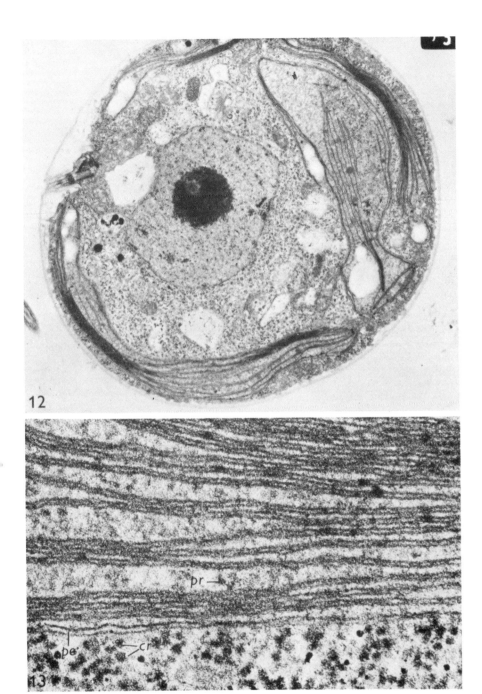

Fig. 14. Section through a typical cell of the mutant *ac-20* grown phototrophically. The lamellar system of the cup-shaped chloroplast is organized largely into 2-disk grana. A large mitochondrial aggregate is evident in the cytoplasm. × 12600.

Fig. 15. A portion of the chloroplast from a cell of *ac-20* showing part of the well developed pyrenoid which is formed under phototrophic growth. × 60000.

Fig. 16. A portion of the *ac-20* mutant cell of Fig. 15 at high magnification showing ribosomes in the cytoplasm and their relative absence in the chloroplast. × 150000.

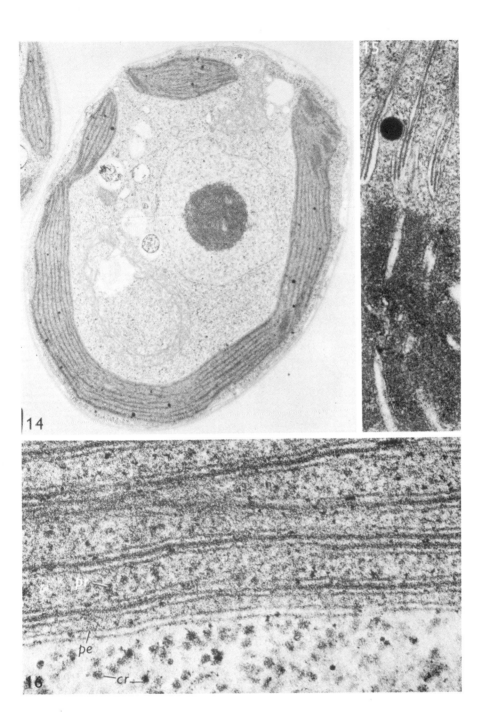

Fig. 17. Section through a typical cell of the mutant *ac-20* grown mixotrophically. The extensive lamellar system in the chloroplast exists either as single, unpaired disks or as giant grana. No pyrenoid is formed. × 12600.

Fig. 18. A section through 2 giant grana in the chloroplast of a mixotrophically grown cell of *ac-20*. × 70000.

Fig. 19. A portion of the *ac-20* mutant cell of Fig. 17 at high magnification showing ribosomes in the cytoplasm and their relative absence in the chloroplast. × 150000.

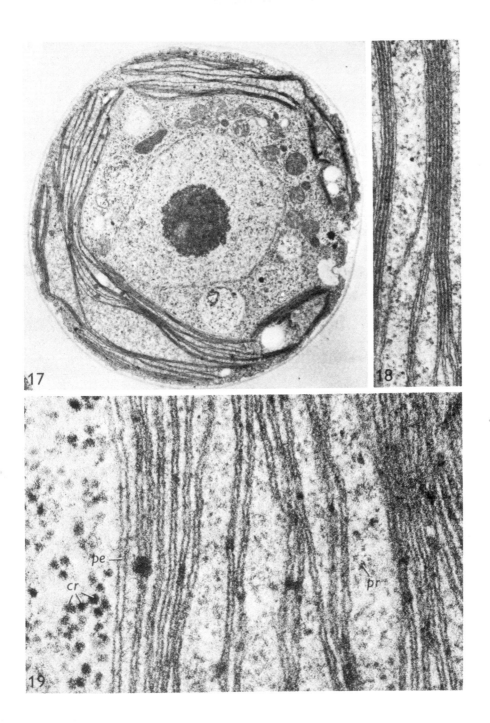

Fig. 20. Section through a typical cell of the mutant *cr-1* grown phototrophically. The lamellar system of the cup-shaped chloroplast is organized largely into 2-disk grana. × 12600.

Fig. 21. A portion of the chloroplast from a cell of *cr-1* showing the abnormally organized pyrenoid formed under phototrophic growth. × 60000.

Fig. 22. A portion of the *cr-1* mutant cell of Fig. 17 at high magnification showing ribosomes in both the cytoplasm and the chloroplast. × 150000.

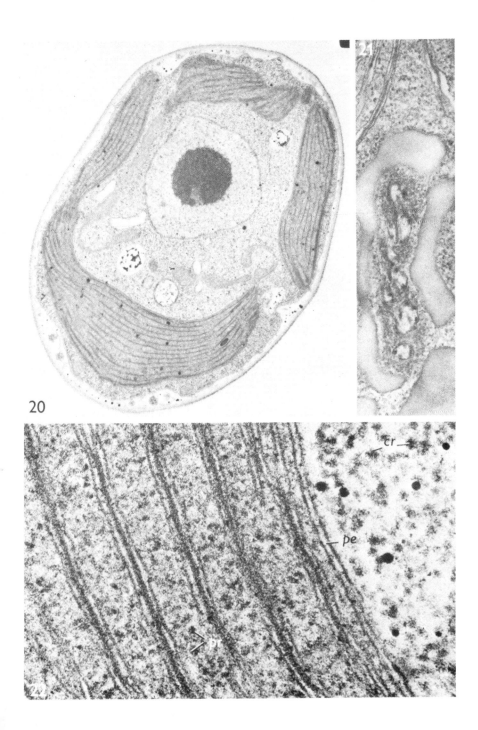

20

cr

pe

pr

Fig. 23. Section through a typical cell of the mutant *cr-1* grown mixotrophically. The extensive lamellar system of the chloroplast exists as either single unpaired disks or as giant grana. No pyrenoid is formed under these growth conditions. × 12600.

Fig. 24. A portion of the *cr-1* mutant cell of Fig. 23 at high magnification showing ribosomes in both the cytoplasm and the chloroplast. × 150000.

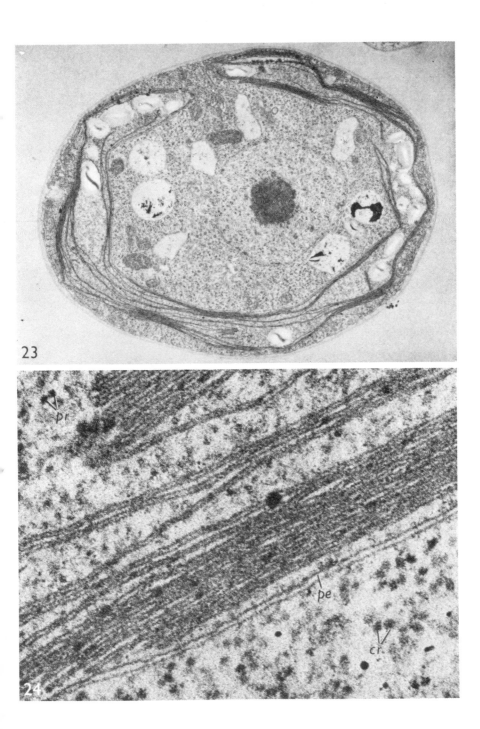

Effects of Inhibitors of Macromolecular
Synthesis on Algae Structure and Metabolism

SELECTIVE INHIBITION BY CYCLOHEXIMIDE OF RIBOSOMAL RNA SYNTHESIS IN CHLORELLA

F. WANKA and P. J. A. SCHRAUWEN

SUMMARY

The net increase of RNA in Chlorella was strongly inhibited by cycloheximide. Incorporation during 2.5 h of [14C]uracil into rRNA was depressed for more than 80 %, while tRNA synthesis was hardly affected. The results suggest that rRNA biosynthesis depends on continuing formation of specific proteins.

INTRODUCTION

Studies on the regulation of RNA synthesis in eucaryotes are primarily dealing with mRNA, while little attention is usually given to tRNA and rRNA. A significant alteration of the rRNA/tRNA ratio was observed during the cell cycle of Chlorella synchronized by alternating periods of light and dark[1]. The regulatory processes controlling the synthesis of the 2 types of RNA are not yet known. A selective inhibition by cycloheximide of chromosomal DNA replication was found recently, suggesting a requirement for simultaneous formation of specific proteins[2-4]. In the present investigation evidence is provided for a similar dependence of rRNA synthesis on continuing protein formation.

EXPERIMENTAL

The experiments were carried out with synchronous cultures of Chlorella (strain 211/8b, Göttingen), grown axenically under photoautotrophic conditions. Cell synchrony was induced by alternating periods of 16 h light and 8 h dark[5]. During the first 10 h after the beginning of the light period the cellular RNA content increases by about 400 % in the absence of any measurable DNA formation[1,6]. The strong RNA increase during 3 h, indicated by the rise of optical density shown in Table I, is depressed by about 85 % in the presence of 15 μM cycloheximide.

RESULTS AND DISCUSSION

The effect of cycloheximide on rRNA synthesis and incorporation of [14C] uracil into various RNA fractions separated by chromatography on a methylated albumin kieselguhr column is shown in Fig. 1. The total yield of nucleic acids ob-

TABLE I

INHIBITION OP NUCLEIC ACID AND PROTEIN SYNTHESIS BY CYCLOHEXIMIDE

Two aliquots for chemical analyses were taken 3 h after the beginning of the light period. The remaining culture was divided into 2 and cycloheximide at a final concentration of 15 μM was added to one of them. Aliquots for the final determinations were taken 3 h later. The cells were collected by centrifugation for 2 min at 2000 $\times g$ and subjected to a preliminary extraction to remove acid and lipid soluble substances. Nucleic acids were then extracted by hydrolysis with HClO$_4$ (ref. 7) and the absorbance was measured at 260 nm. Protein was extracted from the cell residue with 1 M NaOH[5] and determined by the method of LOWRY et al.[8].

Sample	Cycloheximide (μM)	Content per ml of cell suspension	
		RNA (A unit)	Protein (μg)
Initial	—	0.139	64
Final	—	0.258	102
Final	15	0.152	64

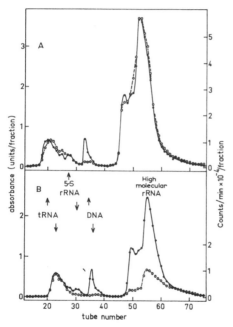

Fig. 1. The effect of cycloheximide on the labeling of various nucleic acid components. A synchronous culture (1.2 · 10^6 cells/ml) was divided into 2 parts of 600 ml 3 h after the beginning of the light period. Part A and B obtained 10 μC [2-^{14}C]uracil each (spec. act. 50 C/mole; New England Nuclear Corp.). To part B, in addition, cycloheximide was added at a final concentration of 15 μM. After 2.5 h the nucleic acids were extracted and deproteinized as described previously[8], and separated by chromatography on methylated albumin kieselguhr columns[10]. Both columns were supplied from the same NaCl gradient with a linear concentration increase of 0.21 mole/100 ml. Fractions of 5.5 ml were collected and the absorbance (●–●) was determined at 260 nm. The radioactivity (O- - -O) was counted in a Packard Tri-Carb scintillation spectrometer, Model 3375, using a dioxane based scintillation fluid[9].

tained from 600 ml of an untreated Chlorella culture was 49 A units while only 31 A units were obtained from an equal culture which was treated with cycloheximide for the last 2.5 h of growth. The difference is almost exclusively due to a lower content of rRNA. Correspondingly, the ^{14}C incorporation into both, 5-S and high molecular rRNA was inhibited for more than 85 %, while labeling of tRNA was only about 10–20 % less in the presence of cycloheximide. This small decrease is in agreement with the general decline of growth and metabolism in the absence of protein formation. Little change was also observed in the high molecular RNA fractions eluted after the major rRNA peak. This is due, at least in part, to the presence of high molecular rRNA precursors, which have been shown to become preferentially labeled for some time in the presence of cycloheximide in yeast[11]. Apparently, like in other eucaryotic cells[12,13], synthesis and processing of ribosomal precursor RNA are both prevented in the presence of cycloheximide. Degradation of unprocessed precursor would produce large amounts of highly labeled, polydisperse break-down products detectable in eluates which do not contain stable nucleic acid components. However, such fractions, eluted before the tRNA and between DNA and high molecular rRNA contain about 60 % less label in cycloheximide treated cells than in the control. For example 614–815 counts/min per fraction were found in Tubes 37–42 of the untreated culture, while only 254–333 counts/min were determined in the corresponding fractions (Tubes 40–45 of Fig. 1) after a cycloheximide treatment. A marked reduction by cycloheximide of the radioactivity was also noticed in the highly labeled fractions eluted after the tRNA peak. Altogether, synthesis and degradation of rapidly labeled RNA seems to be significantly depressed by cycloheximide.

The selective cessation of the increase of both, 5-S and high molecular rRNA in the presence of cycloheximide was also found at other periods of the cell cycle, suggesting a requirement for continuing formation of specific proteins. It seems unlikely that these are enzymes involved in the biosynthesis of nucleic acids, because of the stable levels of such enzyme activities during the cycloheximide treatment[13,14]. It rather might be hypothesized that transcription of the genes of 5-S and high molecular rRNA precursor is activated by a particular class of ribosomal proteins. During inhibition of protein synthesis these would be rapidly depleted by becoming incorporated into maturing ribosomes. The particular dependence on the availability of such proteins might also account for the changing rRNA/tRNA ratio during the cell cycle of Chlorella[1].

ACKNOWLEDGEMENTS

We thank Mr. J. M. A. Aelen for technical assistance. The investigations were supported in part by the Netherlands Foundation for Chemical Research (SON) with financial aid of the Netherlands Organization for the Advancement of Pure Research (ZWO).

REFERENCES

1 F. ENÖCKEL, Z. Pflanzenphysiol., 58 (1968) 457.
2 F. WANKA AND J. MOORS, Biochem. Biophys. Res. Commun., 41 (1970) 85.
3 F. WANKA, J. MOORS AND F. C. N. M. KRIJZER, Biochim. Biophys. Acta, submitted for publication.

4 L. I. GROSSMAN, E. S. GOLDRING AND J. MARMUR *J. Mol. Biol.*, 46 (1969) 367.
5 F. WANKA, *Arch. Mikrobiol.*, 52 (1965) 305.
6 F. WANKA, *Ber. Dtsch. Bot. Ges.*, 75 (1962) 457.
7 F. WANKA, *Planta*, 58 (1962) 594.
8 O. II. LOWRY, N. J. ROSEBROUGH, A. L. FARR AND R. J. RANDALL, *J. Biol. Chem.*, 193 (1951) 265.
9 F. WANKA, H. F. P. JOOSTEN AND W. DE GRIP, *Arch. Mikrobiol.*, 75 (1970) 25.
10 J. D. MANDELL AND A. D. HERSHEY, *Anal. Biochem.*, 1 (1960) 66.
11 S. E. DE KLOET, *Biochen. J.*, 99 (1965) 566.
12 M. WILLEMS, M. PENMAN AND S. PENMAN, *J. Cell Biol.*, 41 (1969) 177.
13 M. MURAMATSU, N. SHIMADA AND T. HIGASHINAKAGAWA, *J. Mol. Biol.*, 53 (1970) 91.
14 F. WANKA AND C. L. M. POELS, *Eur. J. Biochem.*, 9 (1969) 478.
15 O. TH. SCHÖNHERR AND F. WANKA, *Biochim. Biophys. Acta*, 232 (1971) 83.

How methionine and glutamine prevent inhibition of growth by methionine sulfoximine

FREDERICK MEINS, JR. and MARC L. ABRAMS

SUMMARY

Methionine sulfoximine, a potent inhibitor of glutamine synthetase, inhibits growth of *Chlorella vulgaris*. Growth inhibition is prevented by adding L-glutamine or L-methionine to the growth medium. We show that glutamine and methionine act by blocking methionine sulfoximine uptake into the cell.

Methionine sulfoximine induces epileptogenic convulsions in various mammals[1-3] and inhibits the growth of microorganisms[4,5]. Glutamine as well as methionine prevent the toxic effect of methionine sulfoximine when added to the culture medium of microorganisms or when administered to animals at the same time or shortly before methionine sulfoximine[3,5-8]. Since methionine sulfoximine is a potent inhibitor of glutamine synthetase[9], and the effects of methionine sulfoximine are prevented by glutamine, it has been suggested that this compound acts by lowering cellular glutamine concentrations. This interpretation does not account for the effect of methionine since methionine does not prevent or reverse inhibition of glutamine synthetase activity by methionine sulfoximine *in vitro*[10] and is not obviously related to glutamine metabolism. Lamar and Sellinger[9] have shown that methionine inhibits uptake of [^{14}C] methionine sulfoximine into rat brain suggesting that methionine prevents seizures by blocking the entry of methionine sulfoximine into brain cells. In this report we show that both glutamine and methionine prevent the toxic effects of methionine sulfoximine in the unicellular alga, *Chlorella vulgaris*, by blocking methionine sulfoximine uptake into the cell.

We verified Braun's[5] finding that methionine sulfoximine inhibits growth of *Chlorella*. Methionine sulfoximine at a concentration of $2 \cdot 10^{-5}$ M inhibited growth by

50%; 10^{-3} M methionine sulfoximine inhibited growth completely. Growth inhibition was decreased by adding glutamine or methionine to the culture medium (Fig. 1). Since the effect of glutamine or methionine could be prevented by increasing the concentration of methionine sulfoximine, we concluded that these amino acids are competitive inhibitors of methionine sulfoximine action.

Experiments were performed to investigate the relationship of methionine sulfoximine uptake to growth inhibition. [^{14}C]Methionine sulfoximine uptake by *Chlorella* followed Michaelis–Menten kinetics suggesting that this amino acid is transported by a carrier mechanism[12] (Fig. 2). The apparent K_m for uptake, $5 \cdot 10^{-5}$ M, was within experimental error equal to the concentration of methionine sulfoximine required for 50% inhibition of growth. These results indicate that methionine sulfoximine uptake can be the limiting step in growth inhibition. Further support for this conclusion was obtained by comparing [^{14}C]methionine sulfoximine uptake in methionine sulfoximine-resistant and -sensitive strains of *Chlorella*. A resistant strain was isolated that is unaffected by 10^{-3} M methionine sulfoximine which completely inhibits growth of the wild type. This strain took up [^{14}C]methionine sulfoximine at 1/10 the rate of the wild type indicating that methionine sulfoximine sensitivity is correlated with ability to take up methionine sulfoximine (Fig. 3).

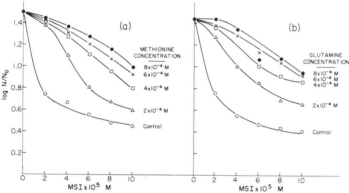

Fig. 1. Prevention of methionine sulfoximine inhibition of growth by methionine and glutamine. Growth experiments followed the method of Braun[11]. Data expressed as the log of cell number at 96 h/initial cell number. Initial cell number = 10^7. a. Increasing concentrations of methionine. b. Increasing concentrations of glutamine. MSI = methionine sulfoximine.

To establish that glutamine and methionine block methionine sulfoximine we compared [^{14}C]methionine sulfoximine uptake in the presence and absence of these amino acids. Both glutamine and methionine inhibited uptake at concentrations that greatly reduce methionine sulfoximine inhibition of growth (Fig. 2). We found that the apparent K_m for uptake was altered, and maximum rate was unchanged indicating that glutamine and methionine are competitive inhibitors.

The fact that we observed competitive inhibition of uptake suggested that glutamine.

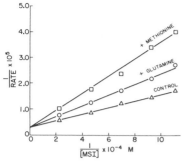

Fig. 2. Effect of methionine and glutamine on uptake of [^{14}C] methionine sulfoximine. Exponentially growing cells were washed twice in an original volume of Braun's[11] medium and incubated for 2 h at 25 °C. Labeling with ^{14}C-labeled amino acids was carried out at 25 °C in 4.0 ml of Braun's medium containing 10^8 washed cells. The incubation mixture in 50-ml Erlenmeyer flasks was shaken at 120 rev./min. Uptake was stopped by filtering the incubation mixture through 0.22 μm pore size membrane filters (Millipore). The filters were then washed with 20 ml of unlabeled amino acid solution (100 mg/1), dried and radioactivity measured in a liquid scintillation spectrometer using a toluene–fluor. L-[^{14}C] Methionine DL-sulfoximine used in uptake experiments was synthesized from L-[^{14}C] methionine using the method of Bentley et al.[13]. Initial rates of uptake were measured and expressed as cpm/10^8 cells per 2 h. Data are presented in double-reciprocal plots. △——△, control; ○——○, 10^{-4} M glutamine added; □——□, 10^{-4} M methionine added.

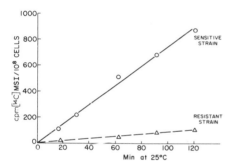

Fig. 3. Comparison of [^{14}C] methionine sulfoximine uptake into resistant and sensitive strains of *Chlorella*. Uptake expressed as cpm [^{14}C] methionine sulfoximine in 10^8 *Chlorella* cells. The concentration of methionine sulfoximine is 10^{-5} M. ○——○, sensitive strain; △ – – △, resistant strain.

methionine and methionine sulfoximine compete for a common carrier. To test this hypothesis, uptake of [^{14}C] glutamine, [^{14}C] methionine and [^{14}C] methionine sulfoximine was measured in the presence and absence of the unlabeled amino acids. We found competitive cross-inhibition of uptake which fulfills the major requirement for transport of several solutes by a common carrier[14]. Another indication of a common carrier is that the K_i for cross-inhibition roughly equals K_m for uptake. This relationship was obeyed by the pairs methionine–methionine sulfoximine and methionine–glutamine, but not by glutamine–methionine sulfoximine (Table I).

Inequality of K_m and K_i for a pair of amino acids does not exclude transport by

TABLE I

K_m AND K_i VALUES FOR COMPETITIVE CROSS-INHIBITION OF METHIONINE, GLUTAMINE, AND METHIONINE SULFOXIMINE UPTAKE

K_i for competitive inhibition and K_m values calculated graphically from Lineweaver–Burk plots of uptake data[17]. Initial rates of amino acid uptake were used in calculations.

Amino acid	K_m uptake × 10^5 (M)	$K_i × 10^5$ (M)		
		Methionine uptake	Glutamine uptake	Methionine sulfoximine uptake
Methionine	1.7	–	2.2	6.0
Glutamine	2.1	4.8	–	15.0
Methionine sulfoximine	5.0	2.0	21.0	–

a common carrier. This behavior is observed when amino acids are transported by several parallel transport systems, e.g. the A and L systems for neutral amino acids in Ehrlich cells[14]. We have found that methionine sulfoximine inhibition of growth reaches a plateau of 73% in the range $8 \cdot 10^{-5} - 10 \cdot 10^{-5}$ M (Fig. 1). Inhibition is complete at 10^{-3} M suggesting that methionine sulfoximine is transported by two systems, one with high affinity, the other with low affinity. Since methionine and glutamine at high concentrations completely prevent methionine sulfoximine toxicity, it is likely that both carrier systems transport the three amino acids. Therefore, we tentatively conclude that inhibition of methionine sulfoximine uptake results from a competition of methionine, glutamine, and methionine sulfoximine for one or more common carriers.

Our results indicate that: (1) methionine sulfoximine uptake can be the limiting step in growth inhibition; and (2) methionine and glutamine competitively inhibit both methionine sulfoximine toxicity and uptake to about the same extent in the same concentration range. Therefore, we concluded that methionine and glutamine prevent the toxic effects of methionine sulfoximine by blocking uptake into the cell. This conclusion is significant in that it shows that the protective effect of glutamine and methionine results from inhibition of methionine sulfoximine transport and not from reversal of glutamine synthetase inhibition or a metabolic by-pass of this enzyme. This interpretation is consistent with the finding in other systems that methionine and glutamine are not effective when administered after methionine sulfoximine[3,7].

Several studies show that lowering of brain glutamine levels and induction of convulsions are independent effects of methionine sulfoximine[15,16]. This suggests that inhibition of glutamine synthetase is not the cause of convulsions. We have found that methionine sulfoximine is a competitive inhibitor of glutamine uptake. Thus, induction of convulsions in mammals could involve inhibition of glutamine transport.

We thank Dr Arthur Pardee and Dr Armin Braun for their helpful suggestions and Mary Moore for expert technical assistance. This study was supported by grants from the National Science Foundation and The Merck Company Foundation. M.L.A. was a participant in the N.S.F. Undergraduate Research Program.

REFERENCES

1　G.W. Newen, T.C. Erikson, W.E. Gilson, S.N. Gershoff and C.A. Elvehjem, *J. Am. Med. Assoc.*, 135 (1947) 760.
2　M. Proler and P. Kellaway, *Epilepsia*, 3 (1962) 117.
3　S.N. Gershoff and C.A. Elvehjem, *J. Nutr.*, 45 (1951) 451.
4　J.G. Heathcote, *Nature*, 164 (1949) 439.
5　A.C Braun, *Phytopathology*, 45 (1955) 659.
6　J.G. Heathcote and J. Pace, *Nature*, 166 (1950) 353.
7　E.L. Peters and D.B. Tower, *J. Neurochem.*, 5 (1959) 80.
8　G W Newen and W.W. Carman. *Fed. Proc.*, 9 (1950) 209.
9　**C. Lamar, Jr. and O.Z. Sellinger, *Biochem. Pharmacol,* 14 (1965) 489.**
10　O.Z. Sellinger and R. Weiler, *Biochem. Pharmacol.*, 12 (1963) 989.
11　A.C. Braun, *Proc. Natl. Acad. Sci. U.S.*, 36 (1950) 423.
12　H.N. Christensen, *Biological Transport*, W.A. Benjamin, New York, 1962.
13　H.R. Bentley, E.E. McDermott and J.K. Whitehead, *Nature*, 165 (1950) 735.
14　H.N. Christensen, *Adv. Enzymol.*, 32 (1969) 1.
15　J. Folbergrova, *Physiol. Bohemoslov.*, 13 (1964) 21.
16　C. Lamar, *Biochem. Pharmacol.*, 17 (1968) 636.
17　M. Dixon and E.C. Webb, *Enzymes*, Longmans, Green and Co., London, 2nd edn, 1964, p. 325.

THE EFFECTS OF INHIBITORS OF RNA AND PROTEIN SYNTHESIS ON CHLOROPLAST STRUCTURE AND FUNCTION IN WILD-TYPE CHLAMYDOMONAS REINHARDI

URSULA W. GOODENOUGH

ABSTRACT

Wild-type cells of the unicellular green alga *Chlamydomonas reinhardi* have been grown for several generations in the presence of rifampicin, an inhibitor of chloroplast DNA-dependent RNA polymerase, spectinomycin and chloramphenicol, two inhibitors of protein synthesis on chloroplast ribosomes, and cycloheximide, an inhibitor of protein synthesis on cytoplasmic ribosomes. The effects of cycloheximide are complex, and it is concluded that this inhibitor cannot give meaningful information about the cytoplasmic control over the synthesis of chloroplast components in long-term experiments with *C. reinhardi*. In the presence of acetate and at the appropriate concentrations, the three inhibitors of chloroplast protein synthesis retard growth rates only slightly and do not affect the synthesis of chlorophyll; however, photosynthetic rates are reduced fourfold after several generations of growth. Each inhibitor produces a similar pattern of lesions in the organization of chloroplast membranes. Only rifampicin prevents the production of chloroplast ribosomes.

INTRODUCTION

The unicellular green alga, *Chlamydomonas reinhardi*, possesses a single large chloroplast that contains DNA and 68S, bacterial-like ribosomes (for review, see reference 22). Investigations in this laboratory have focused on the role that this protein-synthesizing apparatus plays in constructing the structural and functional components of the chloroplast (1, 10, 11, 17, 23, 32–35). The present paper reports a study of the structural and functional lesions produced when wild-type *C. reinhardi* cells are grown for several generations in the presence of antibiotics that specifically inhibit transcription or translation, and hence protein synthesis, in the chloroplast. The antibiotics used are rifampicin, an inhibitor of the chloroplast's DNA-dependent RNA polymerase (32), and

chloramphenicol and spectinomycin, two inhibitors of chloroplast protein synthesis (5, 6, 15). Cycloheximide, an inhibitor of protein synthesis in the cytoplasm (15, 30), is also used in an attempt to estimate the cytoplasm's contribution to the production of a normal chloroplast.

Long-term growth experiments that utilize antibiotics must be performed and interpreted with considerable caution. First, a concentration of antibiotic must be chosen that is high enough to inhibit RNA or protein synthesis but not high enough to be generally toxic to the cells during the course of the 48–72-hr growth periods required to dilute out existing chloroplast components by cell division. In particular, concentrations of antibiotic must be used that do not affect mitochondrial protein synthesis, for one assumes that *C. reinhardi* cannot survive without functional mitochondria (33). It is important to establish that the cells are not dying, since the loss of cellular activities and of structural integrity proceeds in a most nonspecific manner in moribund cells. Second, it has been found that when wild-type cells are exposed to an antibiotic for long periods of time, they sometimes make an adjustment to the exposure. For example, after 24 hr of growth in 3 μg/ml of spectinomycin, cells acquire a resistance to this concentration of the antibiotic (the origin and nature of this resistance has not yet been determined). One cannot, however, go to higher concentrations of spectinomycin because these are toxic to the cells during the first 24 hr of the experiment.

Despite these complications, long-term growth experiments can provide certain kinds of information—notably, information as to the effects of the inhibitors on chloroplast ribosome levels and chloroplast fine structure—that cannot be obtained in short-term experiments with wild-type cells (1).

MATERIALS AND METHODS

Culture of the Organisms

The plus and minus mating types of the wild-type strain 137c of *C. reinhardi* were used with comparable results. The *spa-2* mutant strain was used in certain experiments; its properties are described in the Results section.

For experiments investigating the effects of spectinomycin, chloramphenicol, or cycloheximide, cells were grown overnight in 300 ml volumes of Tris-acetate-phosphate (TAP) medium (13) contained in 500-ml Erlenmeyer flasks. These were agitated on rotary shakers in the light (2500 lux) at 26°C. The following morning, the culture was diluted twofold and divided between two 500-ml flasks, one serving as the control. All procedures were performed under sterile conditions. The following stock solutions (in water) of antibiotics were prepared the day of the experiment and sterilized by passage through a Millipore filter (Millipore Corp., Bedford, Mass.): spectinomycin (a gift from Upjohn Co., Kalamazoo, Mich.), 3 mg/ml; chloramphenicol (Sigma Chemical Co., St. Louis, Mo.), 2.5 mg/ml; and cycloheximide (Sigma Chemical Co.), 3 mg/ml. Appropriate aliquots of these were added to the experimental flask to give the final antibiotic concentrations cited in the text. Growth was continued for an additional 30–96 hr; where growth was continued for more than 48 hr, cells were transferred to fresh medium containing fresh antibiotic. Cell number during the

growth period was determined with the aid of a hemacytometer.

The above procedures were modified in experiments investigating the effects of rifampicin, for the reasons given in the Results section. Two 300-ml volumes of TAP medium were each inoculated in the dark with 10 ml of a culture of wild-type cells whose growth had been synchronized (19) with a light-dark cycle for three generations. The cells in the inoculum were in the final hour of the dark cycle. To the experimental flask was also added a sample of a rifampicin stock solution prepared by adding 125 mg rifampicin (Mann Research Labs Inc., New York) to 5 ml of sterile 0.01 M KH_2PO_4, pH 4.5, and agitating in the dark with a magnetic stirrer for 8 hr to insure proper solution of the antibiotic. Both control and experimental flasks were maintained in darkness throughout the growth period. The cultures were then exposed to light for 2.5 hr before harvesting.

Measurement of Photosynthetic Parameters

Total chlorophyll and chlorophylls *a* and *b* were determined by a modification (2) of the method of Mackinney (26). Photosynthetic CO_2 fixation was measured as described previously (8).

Electron Microscopy

Cells were fixed and embedded as previously described (10, 18), and thin sections were examined with a Hitachi HU-11C electron microscope. Procedures for making ribosome counts from electron micrographs have been described previously (10).

RESULTS

Control Cells

The growth rate, chlorophyll content and chlorphyll *a:b* ratio, rate of photosynthetic CO_2 fixation, chloroplast ribosome content, and chloroplast membrane organization of wild-type *C. reinhardi* cells grown mixotrophically—in the light in the presence of a source of fixed carbon (acetate)—have been described in several other publications from this laboratory (3, 8–10, 21, 24, 35). Representative values for most of these parameters are given in Table I.

In experiments investigating the effects of rifampicin, control cells were grown heterotrophically—in the dark in the presence of acetate—for 3–4 days and were then exposed to light for 2.5 hr. The rationale for these growth conditions will be presented shortly. Control cells grown under such conditions have almost normal levels of chloro-

TABLE I

Chlorophyll Content, Photosynthetic Capacity, and Ribosome Levels in C. reinhardi Cells Grown in the Absence and in the Presence of Antibodies

Antibiotic	Strain, growth condition	Mean doubling time	Divisions before harvest	µg Chl/10^4 cells	Chl a:b	µmoles CO_2 fixed/hr				Chloroplast ribosomes		Cytoplasmic ribosomes
		hr	No.			·mg Chl	% control	·10^8 cells	% control	No. per stroma weight	% control	No. per ground substance weight
Control	wild-type, mixotrophic	11	5.5	3.6	2.3	121		0.436		200		405
Control	wild-type, heterotrophic + 2.5 hr light	14	4	2.8	2.5	124		0.356		271		538
Control	spa-2, 10 µg/ml spectinomycin, mixotrophic	12	ca. 4	*	2.1	103		*		209		585
Rifampicin (250 µg/ml)	wild-type, heterotrophic + 2.5 hr light	20	3.5	1.3	2.1	25	(20%)	0.032	(9%)	43	(16%)	440
Spectinomycin (25 µg/ml)	spa-2, mixotrophic	12	ca. 10	*	2.0	29	(28%)	*		238	(114%)	610
Chloramphenicol (100 µg/ml)	wild-type, mixotrophic	17	3	3.0	2.2	36	(30%)	0.108	(25%)	208	(104%)	450
Cycloheximide (1 µg/ml)	wild-type, mixotrophic	48	1	7.6	2.8	126	(104%)	0.955	(220%)	368	(194%)	645

* Not possible to obtain data; see text.

phyll and rates of CO_2 fixation (Table I), and most of them exhibit the normal pattern of chloroplast membrane organization found in mixotrophically grown cells, namely, anastomosing stacks of from 2 to 10 thylakoids (9). Perhaps 20% of the cells in such cultures, however, exhibit the distinct chloroplast phenotype illustrated in Fig. 1. The chloroplast membranes are organized into what appear to be short, segmented thylakoids that frequently fold back on themselves. Cells that contain such membrane aggregates do not contain normal thylakoid stacks; they thus appear to represent a distinct population of cells within the culture. Because of their morphological similarity to "yellow" strains of *C. reinhardi* (16, 17, 29), it is likely that these cells represent a "yellow" clone within our wild-type stock. Despite their anomalous membrane organization, such cells possess abundant chloroplast ribosomes (Fig. 1); indeed, as seen in Table I, heterotrophically grown wild-type cells of *C. reinhardi* exhibit higher levels of chloroplast ribosomes than do their light-grown counterparts.

In experiments investigating the effects of spectinomycin it was found, as mentioned in the Introduction, that wild-type cells were killed by a concentration of 5 µg/ml of spectinomycin but acquired a resistance to 3 µg/ml of spectinomycin. To obtain meaningful data with this antibiotic it was therefore necessary to use a strain that was already resistant to low spectinomycin levels and that would tolerate a wider range of concentrations of the antibiotic. These requirements were met by the *spa-2* strain of *C. reinhardi*, recently isolated from the wild-type strain in this laboratory by J. J. Armstrong. Cells of this strain can be grown in liquid TAP cultures in the presence of 10 µg/ml of spectinomycin, and their growth rates, chloroplast fine structure, and photosynthetic capacities are indistinguishable from those of the wild-type cells (Table I). Therefore, the control cells in spectinomycin experiments are *spa-2* cells grown on 10

µg/ml of spectinomycin, and the experimental cells, to be described shortly, are *spa-2* cells grown on 25 µg/ml of spectinomycin. Concentrations of 50–100 µg/ml of spectinomycin are required to kill *spa-2* cells on liquid TAP medium.[1]

The major drawback to working with *spa-2* cells is that they become paralyzed during growth in liquid cultures, whether or not spectinomycin is present in the medium. It can be seen with the electron microscope that normally constructed but very short flagella are formed by such cells. Because they cannot swim, daughter cells tend to remain within the mother wall after cell division and, as a result, the culture comes to contain large clumps of cells that cannot be accurately counted with a hemacytometer. Growth rates can therefore only be estimated approximately (Table I), and it is impossible to give reliable data on chlorophyll content per cell. It is not known whether there is any relationship between the strain's paralysis and its resistance to low levels of spectinomycin.

Cells Grown in the Presence of Rifampicin

Rifampicin interacts with the bacterial (14), and apparently also with the *C. reinhardi* chloroplast (32) DNA-dependent RNA polymerase in such a way as to prevent the transcription of bacterial and chloroplast DNA. The antibiotic interacts with the polymerase only when the polymerase is not associated with its DNA template (36). In order to see a maximal effect of rifampicin on chloroplast DNA transcription it is, therefore,

---

[1] The genetics of the *spa-2* strain and the spectinomycin-binding properties of *spa-2* chloroplast ribosomes will be the subject of separate communications from this laboratory (S. J. Surzycki and W. Burton, manuscripts in preparation), which demonstrate that the *spa-2* mutation is inherited in a Mendelian fashion and that 68S ribosomes isolated from *spa-2* cells bind spectinomycin in the same way as wild-type 68S ribosomes.

FIGURE 1 Portion of a cell from the wild-type strain of *C. reinhardi* grown heterotrophically in the dark for 4 days and exposed to light for 2.5 hr before being fixed. Chloroplast ribosomes (arrowhead are at the normal wild-type level (Table I). Chloroplast membranes exist largely as segmented stacks of thylakoids, suggesting that the cell derives from a "yellow" clone (16, 17, 29) within the wild-type population. Perhaps 20% of the cells in the sample exhibit this phenotype; the remainder are indistinguishable from mixotrophically grown wild-type cells illustrated in other publications (7, 9, 10, 12, 18). × 95,000.

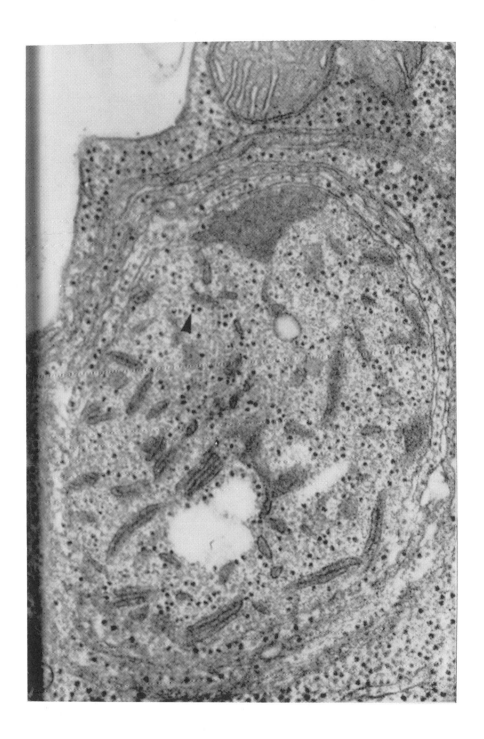

necessary to start with cells whose chloroplast DNA-dependent RNA polymerase is in the "unattached" state. This is accomplished by inoculating the culture with cells whose growth has been synchronized on a minimal medium by a light-dark cycle (1, 19). Cells are taken from the end of a 12 hr dark cycle; at this time they are engaged in no detectable chloroplast RNA synthesis (Surzycki and Hastings, manuscript in preparation) and thus the chloroplast DNA-dependent RNA polymerase is free to interact with rifampicin.[2]

Two other features of long-term growth experiments with rifampicin should be noted. First, the experiments are carried out in the dark under heterotrophic growth conditions because the antibiotic, a bright red-orange in color, has the effect of screening out a great deal of the light available to a control culture; moreover, light appears to inactivate rifampicin with time. Second, the heterotrophically grown cultures are exposed to 2.5 hr of light before being harvested. This step is taken because it has been shown (R. P. Levine and D. Graham, unpublished observations) that, after three to four generations of heterotrophic growth, wild-type *C. reinhardi* cells cannot carry out a maximal rate of photosynthesis until after they are exposed to light for about 2.5 hr.

CHLOROPLAST RIBOSOMES: Cells grown heterotrophically in the presence of 250 μg/ml of rifampicin under conditions described above contain only a fraction of the level of chloroplast ribosomes found in heterotrophically grown control cells. This is seen in Fig. 2 and in the chloroplast ribosome counts given in Table I. A slight reduction is also found in counts of cytoplasmic ribosomes in rifampicin-grown cells (Table I); how-

ever, since the value obtained is still within the normal range for cytoplasmic ribosomes (Table I) and since the tight packing of cytoplasmic ribosomes renders them very difficult to count, this observation remains of uncertain significance. Certainly the 18% reduction in levels of cytoplasmic ribosomes, if real, is an effect that is very different from the 84% reduction in levels of chloroplast ribosomes.

On the assumption that rifampicin brings about a 100% cessation of chloroplast ribosomal RNA synthesis (32), and that existing chloroplast ribosomes are lost from the cells by the dilution that accompanies cell division, one can calculate that, after 3.5 generations, cells should possess perhaps 9% of the control levels of chloroplast ribosomes. That they possess 16% of the control levels can be attributed to such considerations as an incomplete atibiotic-enzyme interaction and sampling errors; in other words, the discrepancy between observation and theory does not appear great.

CHLOROPLAST FINE STRUCTURE: Growth in the presence of rifampicin brings about a severe disorganization of chloroplast membranes in wild-type cells. A branching system of membranes fills the chloroplast interior of virtually every cell (Fig. 2); these membrane profiles will be referred to as vesicles since they give a vesiculate appearance in section. A comparison of Figs. 1 and 2 indicates that the vesicles are quite different in morphology from the truncated thylakoids observed in a small portion of the heterotrophically grown control cells.

Although vesicles are the prominent membrane configuration, chloroplasts of rifampicin-grown cells typically contain some normal thylakoids as well. Most commonly, these exist as single, unstacked thylakoids (Fig. 2); they are also found in wide stacks beneath the chloroplast envelope (Fig. 2).

When control wild-type cells are grown in the

---

[2] Synchronized cells taken at the end of a dark cycle were also used by Surzycki in his in vitro experiments with rifampicin, a fact that was inadvertently omitted from his published results (32).

FIGURE 2 Portion of a cell from the wild-type strain of *C. reinhardi* grown heterotrophically in the dark for 4 days in the presence of 250 μg/ml rifampicin and exposed to light for 2.5 hr in the presence of rifampicin before being fixed. Chloroplast ribosomes (arrowheads) are sparse (Table I). Chloroplast membrane is seen as single thylakoids (*T*) and as masses of tubular profiles that appear as vesicles (*V*). A region of DNA (*DNA*) lies next to a pyrenoid (*P*) included in grazing section. Starch (*S*) levels are high. Ribosomes are abundant in the cytoplasm (*C*). The outer membrane of the mitochondrion in this section is absent; this is a fixation artifact which is seen only in occasional cells in the sample and which is unrelated to the effect of the antibiotic. × 68,000.

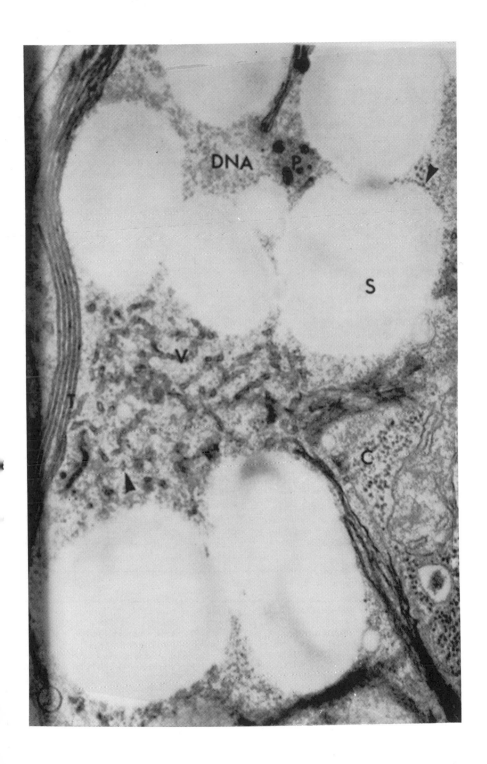

dark, pyrenoid formation is apparently abated: the chloroplasts contain much smaller pyrenoids than do the chloroplasts of light-grown wild-type cells. A similar reduction in pyrenoid size, and hence in the frequency with which pyrenoids are encountered in thin sections, is observed in cells grown in the presence of rifampicin; the antibiotic does not, however, appear to bring about any additional change in the cells's pyrenoid-forming capacity.

Starch production is stimulated by heterotrophic growth conditions in both control and rifampicin-treated cells (see also reference 29).

Growth in rifampicin is without effect on the structure of mitochondria at the concentrations used. The fine structure of other cell organelles (except the chloroplast) is also normal. This statement applies as well to cells grown in the presence of spectinomycin and chloramphenicol.

PHOTOSYNTHETIC CAPACITY: Heterotropically grown wild-type cells typically contain about half the chlorophyll of light-grown cells, a difference that is rapidly erased when the cells are exposed to light. Under heterotrophic conditions, rifampicin-grown cells contain somewhat less chlorophyll than the control cells (Table I), but their ability to form chlorophyll has by no means been blocked by the antibiotic. The ratio of chlorophyll a to chlorophyll b in these cells also lies in the normal range (Table I).

Photosynthetic capacity, measured as the ability of cells to carry out a light-stimulated incorporation of $^{14}CO_2$ into carbohydrate, is greatly reduced by growth in the presence of rifampicin (Table I). When measured on a cell basis, the photosynthetic capacity has been reduced to 9% of the control; the reduction is less great (20%) when calculated on a chlorophyll basis since the cells are chlorophyll deficient.

Cells Grown in the Presence of Spectinomycin

Experiments with spectinomycin were carried out with the *spa-2* strain of *C. reinhardi* grown on 25 µg/ml of spectinomycin; the properties of this strain are described in an earlier section of this paper.

CHLOROPLAST RIBOSOMES: Long-term growth in the presence of spectinomycin has no effect on the ability of cells to form chloroplast ribosomes (Fig. 3 and Table I). Cytoplasmic ribosome formation is also unaffected by spectinomycin (Table I).

CHLOROPLAST FINE STRUCTURE: Growth in spectinomycin produces occasional cells whose chloroplasts contain vesiculate masses that resemble those found in rifampicin-grown cells. Commonly, however, thylakoid formation proceeds normally. The thylakoids that are formed may fuse into normal stacks or into high stacks beneath the chloroplast envelope (Fig. 3), but most characteristically they are found in the unstacked configuration (Fig. 3). The single thylakoids usually lie in the chloroplast interior, each following an independent, meandering course through the chloroplast stroma.

In cells exposed to toxic levels of spectinomycin (5 µg/ml for wild-type and 100 µg/ml for *spa-2*), a very different membrane phenotype is observed. Wide bands of perhaps 20–23 thylakoids aggregate beneath the chloroplast envelope, and the intrathylakoid spaces become irregularly swollen and collapsed so that the whole structure has a "ruffled" appearance.

Pyrenoid formation appears to be disrupted by growth in spectinomycin. The pyrenoids are small and hence are encountered less frequently in section compared to sections of control cells. Incursions of chloroplast stroma are frequently made into the pyrenoid matrix such that the matrix has a mottled appearance and lacks its usual polygonal symmetry (7, 29).

Starch formation suffers no apparent interruption during growth in spectinomycin.

PHOTOSYNTHETIC CAPACITY: Because of the "clumping" phenomenon described earlier, accurate values for the amount of

---

FIGURE 3 Portion of a cell from the *spa-2* strain of *C. reinhardi* grown mixotrophically for 5 days in the presence of 25 µg/ml spectinomycin. Chloroplast ribosomes (arrowhead) are at the normal wild-type level (Table I). Thylakoids stack in wide bands (*B*) beneath the chloroplast envelope and course singly through the stroma. This cell has recently completed mitotic division and exhibits two "dense bodies" (*D*) that characterize the mitotic chloroplast of *C. reinhardi* (7). × 79,000.

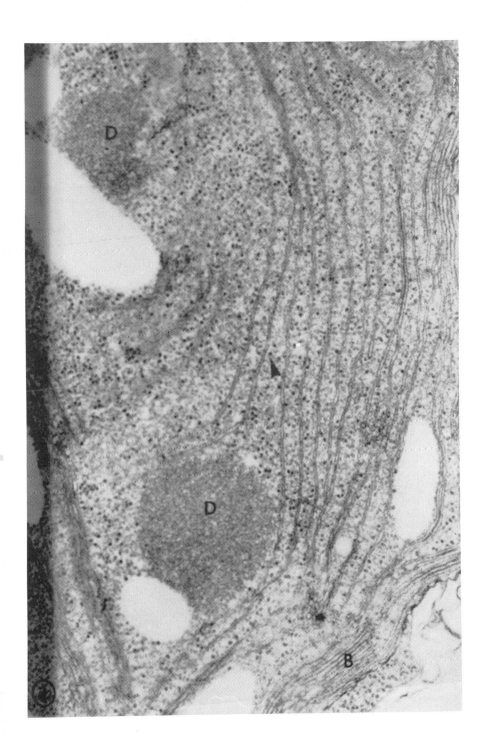

chlorophyll per cell cannot be given for *spa-2* cells grown in the presence of 25 µg/ml spectinomycin. The cultures appear fully green, however, and when wild-type cells are grown in spectinomycin, they contain normal levels of chlorophyll even when the concentrations of spectinomycin are toxic.

Long-term growth in 25 µg/ml of spectinomycin produces *spa-2* cells that fix CO_2 at 28% of the rate of *spa-2* cells grown on 10 µg/ml of spectinomycin (Table I). If the concentration of spectinomycin is raised to 100 µg/ml, cells fix CO_2 at only 4% of the control rate, but such cells are also moribund. Presumably, with sufficiently experimentation, a concentration of spectinomycin between 25 and 100 µg/ml could be found that would exert a greater inhibitory effect on the cells's photosynthetic capacity than does 25 µg/ml without simultaneously killing the cells, but for the purposes of the present experiments it was sufficient that the cells' photosynthetic capacity be inhibited fourfold.

Cells Grown in the Presence of Chloramphenicol

CHLOROPLAST RIBOSOMES: As with spectinomycin, growth on 100 µg/ml of chloramphenicol has no effect on the cells's ability to form chloroplast, or cytoplasmic, ribosomes (Fig. 4 and Table I).

CHLOROPLAST FINE STRUCTURE: In three different experiments, three different degrees of chloroplast membrane disorganization were observed in chloramphenicol-grown cells. In all cases, vesicle formation was prominent (Fig. 4). The vesicles are indistinguishable from those formed by rifampicin-grown cells (Fig. 2) and occasionally by cells in spectinomycin-grown cultures. Considerable variation was found, however, in the degree of thylakoid unstacking that accompanied this vesicle formation; the source of this variation is not understood. In one experiment, considerable unstacking was observed, none was found in a second experiment, and an intermediate level was found in a third. Normal stacks and high stacks beneath the chloroplast envelope are also found in chloramphenicol-grown cells.

Pyrenoid formation is disrupted by growth in chloramphenicol in the same way that it is disrupted in spectinomycin-grown cells (Fig. 5). Starch formation is normal.

PHOTOSYNTHETIC CAPACITY: The ability to synthesize chlorophyll and to produce chlorophylls *a* and *b* in normal proportions is unaffected by chloramphenicol (Table I).

Growth on 100 µg/ml of chloramphenicol for three generations produces a 70% inhibition in CO_2 fixation rates (Table I); theoretically, one would predict an 87% inhibition by dilution concomitant with cell division. The discrepancy between theory and observation may be caused by a choice of too low a concentration of chloramphenicol, by the fact that chloramphenicol apparently begins to lose its inhibitory effects by 48 hr (see following section), or, probably, by a combination of the two effects. A comparable inhibition of photosynthesis was observed in each of the three chloramphenicol experiments performed even though variation in chloroplast structure was observed.

Cells maintained in the Presence of Cycloheximide

Exposure of wild-type *C. reinhardi* cells to low concentrations (1 µg/ml) of cycloheximide produces some bizarre effects. For the first 12 hr, cell division is completely blocked and no change is observed in cell size or in levels of chlorophyll per cell. By 18 hr, an increase in cell size is perceptible. By 24 hr, a few cells have divided (cell number may increase by 30%) and the remaining cells are even larger; chlorophyll per cell values are very high, being in some experiments five times higher than the control. By 48 hr, the cell number of the culture has roughly doubled. Typically, this doubling results from a portion of the cells in the culture undergoing three or four successive mitotic divisions so that clusters of 16 or 32 cells form within greatly distended mother walls; the remaining cells, those that have not yet undergone division, are enormous, perhaps five times the size of normal wild-type cells. The increase in size is accompanied by a dramatic increase in size of all the cells' organelles: with the electron microscope, one observes immense nuclei, a great proliferation of rough endoplasmic reticulum, huge chloroplasts containing giant pyrenoids, and so on. If cells are resuspended in fresh cycloheximide at the end of 24 hr, the same sequence of events is observed.

Since cytoplasmic protein synthesis is reportedly blocked by cycloheximide, such increases in total cell mass were difficult to understand. It was found, however, that the increases could be prevented if cells were grown in both cycloheximide (1 µg/ml)

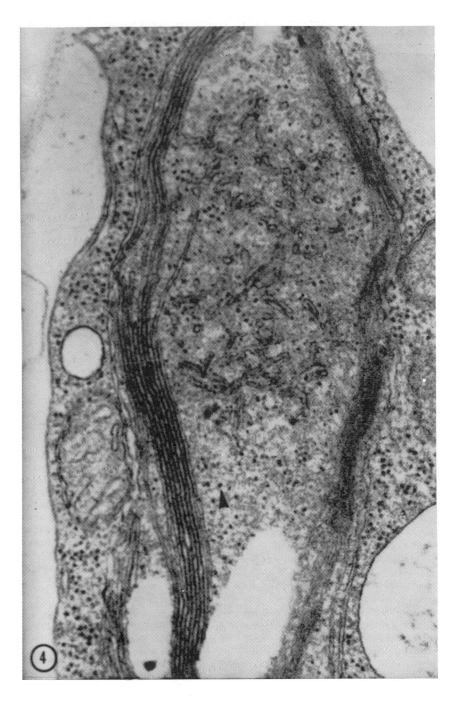

FIGURE 4 Portion of a cell from the wild-type strain of *C. reinhardi* grown mixotrophically for 48 hr in the presence of 100 μg/ml chloramphenicol. Chloroplast ribosomes (arrowhead) are at the normal wild-type level (Table I). Chloroplast membrane is found in wide bands (*B*) beneath the chloroplast envelope and in aggregates of tubular vesicles (*V*). × 97,000.

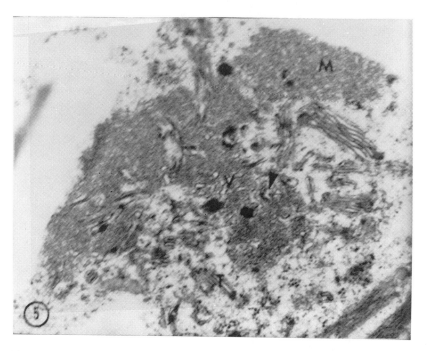

FIGURE 5 Pyrenoid from a wild-type cell of *C. reinhardi* grown mixotrophically for 48 hr in the presence of 100 μg/ml chloramphenicol. Normal pyrenoid structure has been greatly disrupted: the pyrenoidal matrix (*M*) is interrupted by incursions of chloroplast stroma, and most of the normal pyrenoidal tubules (arrowhead) are replaced by vesicles (*V*) and truncated thylakoids (*T*). × 63,000.

and chloramphenicol (100 μg/ml), Such cells undergo no cell division and synthesize no chlorophyll during the first 24 hr of growth. By 48 hr, some cell division begins to occur, and if the culture is allowed to grow for 5 days, it eventually reaches the stationary phase of growth.

Summarizing these observations, it appears that cycloheximide is able to effect a complete block over cytoplasmic protein synthesis for a 12 hr period; a slow, and apparently unregulated, protein synthesis then commences, one that is totally inhibited by chloramphenicol; and finally, a protein synthesis proceeds that is insensitive to both inhibitors. The simplest interpretation of the second "phase" of this sequence is to suggest that messenger RNA's that are normally translated on cytoplasmic ribosomes are able, under the stress of cycloheximide inhibition, to find their way into the chloroplast and be translated on

chloroplast ribosomes. The simplest interpretation for the final "phase" of the process is that, after 48 hr, resistant cells arise in the culture, or the antibiotics lose their inhibitory properties, or both. Other, more complex, interpretations can also be envisaged. The important point is that cycloheximide can be used as a reliable inhibitor of cytoplasmic protein synthesis in *C. reinhardi* only during a 12 hr experiment, and since no cell division occurs during the first 12 hr in cycloheximide, a meaningful growth experiment with cycloheximide cannot be designed for *C. reinhardi*.

The data in Table I indicate that chloroplast ribosomes, cytoplasmic ribosomes, amount of chlorophyll per cell, and photosynthetic capacity are all either unaffected or show an increase in cells that have been exposed to cycloheximide for 48 hr. Chloroplast membrane organization is also normal except that most of the thylakoids are in

152

stacks of two. In view of the most unusual sequence of events that occurs during the 48 hr period, no information can be gleaned from these data as to the role of cytoplasmic protein synthesis in forming a chloroplast in *C. reinhardi*.

DISCUSSION

The experiments reported in this paper establish that wild-type cells of *C. reinhardi* can be grown mixotrophically or heterotrophically for periods of 48–96 hr in the presence of appropriate concentrations of antibiotics that inhibit protein synthesis in the chloroplast. Under these conditions the cells remain viable and their growth rates are comparable to those of control cells (Table I). Their ability to fix CO_2 by photosynthesis, on the other hand, is progressively lost as component(s) of their existing photosynthetic apparatus are diluted by successive cell divisions. Thus it is concluded that protein synthesis in the chloroplast of *C. reinhardi* is required to construct a functional photosynthetic apparatus. The loss of photosynthetic ability is never complete in these experiments, since it was deemed more important to work with healthy cells whose chloroplast protein synthesis is not completely inhibited than to work with fully inhibited cells that might also be dying.

With the exception of chlorophyll, no attempt was made in these experiments to determine which specific components of the photosynthetic apparatus are affected by the antibiotics. Such determinations can only be made in a meaningful way in short-term experiments such as those reported in the accompanying paper (11), in the paper of Armstrong et al. (1), and in the paper of Hoober et al. (16). In long-term experiments, the apparent loss of a given component may be the secondary result of the loss of a first component rather than the direct effect of inhibiting protein synthesis (see, in this regard, the papers of Smillie et al. [31] and Linnane and Stewart [25]). This point is well illustrated by the case of chlorophyll in rifampicin-grown wild-type cells. Such cells consistently contain only about half the chlorophyll of control cells (Table I). However, this deficiency is apparently a secondary result of the highly disordered chloroplast membrane conformations produced by growth in rifampicin, since in short-term experiments rifampicin has no inhibitory effect over chlorophyll synthesis (1).

An inhibition of protein synthesis on chloroplast ribosomes by spectinomycin or chloramphenicol that is sufficient to effect a 70–80% inhibition of photosynthesis has no effect on levels of chloroplast ribosomes (Table I). In the case of the spectinomycin experiments, this was true even after 10 generations of growth in the presence of the antibiotic. It is not known whether the ribosomes that are formed in the presence of these inhibitors possess all the ribosomal proteins in normal amounts, but if any proteins are missing, their absence does not prevent the construction of an intact ribosomal particle. Thus it is concluded that at least most of the proteins that make up chloroplast ribosomes in *C. reinhardi* are synthesized outside the chloroplast, presumably on cytoplasmic ribosomes. Similar conclusions have been reached for the proteins of mitochondrial ribosomes in fungi (4, 20, 27, 28).

The fact that chloroplast ribosomes continue to be made even though chloroplast protein synthesis is inhibited for as long as 10 generations would suggest that the chloroplast DNA-dependent RNA polymerase is also synthesized outside the chloroplast. Direct assays of the enzyme are necessary to confirm this inference.

The experiments with rifampicin reported in this paper demonstrate that a block in chloroplast DNA transcription leads to a dramatic cessation in the production of chloroplast ribosomes (Table I). Surzycki (32) has shown that long-term growth in rifampicin produces a dramatic loss in the cells' 16S and 23S RNA species, and Hoober and Blobel (15) have demonstrated that the 68S ribosomes of *C. reinhardi* contain 16S and 23S RNA's. Taken together, these results indicate that at least the bulk of the 68S ribosomes in *C. reinhardi* can be equated with ribosomes that are located within the chloroplast. They also confirm Surzycki's conclusion (32) that information for the synthesis of chloroplast ribosomal RNA resides exclusively in chloroplast DNA in *C. reinhardi*.

Growth in rifampicin, chloramphenicol, or spectinomycin brings about a similar pattern of membrane disorganization: small vesicles form, thylakoids do not stack properly, and wide membranous bands form beneath the chloroplast envelop. Such membrane conformations are not produced when wild-type *C. reinhardi* cells are grown in the presence of cycloheximide; they do not form when the photosynthetic capacity of cells is curtailed by gene mutation (8, 9, 12) or by growth in the presence of an herbicide such as 3-(3,4-dichlorophenyl)1,1-dimethylurea (19); nor

153

do they form if cells are exposed to toxic concentrations of the inhibitor spectinomycin. It is therefore concluded that the inhibition of chloroplast protein synthesis brings about a disruption in the ability of *C. reinhardi* to form a normally constructed chloroplast; moreover, this disruption produces a specific set of morphological lesions. Similar conclusions have also been reached by Hoober et al. (16) in their study of the effects of chloramphinicol on the re-greening process in the *y-1* mutant strain of *C. reinhardi*.

Summarizing, the results of long-term growth experiments indicate that the chloroplast protein-synthesizing system in *C. reinhardi* is required for the proper organization of chloroplast membranes, proper pyrenoid formation, and the formation of a functional photosynthetic apparatus. In an earlier study of *ac-20*, a mutant strain of *C. reinhardi* that possesses drastically reduced levels of chloroplast ribosomes, the cells were shown to possess rudimentary pyrenoids, vesiculate or unstacked chloroplast membranes, and a defective photosynthetic capacity (10, 23, 35). The present paper shows that an *ac-20* "syndrome" can be simulated in wild-type cells by inhibiting their chloroplast protein synthesis, thus supporting our earlier conclusion (10) that the *ac-20* syndrome is produced as a consequence of the strain's low levels of chloroplast ribosomes.

I am grateful for the interest and advice given throughout this project by Professor R. P. Levine, Dr. S. J. Surzycki, and Miss J. J. Armstrong.

This work was supported by a postdoctoral fellowship from the National Institutes of Health (GM 24306), by the Maria Moors Cabot Foundation for Botanical Research, Harvard University, and by Research Grant GB 1866 from the National Science Foundation to R. P. Levine.

Abbreviated reports of this research have been presented at the Symposium of the Society for Experimental Biology, London, 1969 (33), and at the Symposium on the Biogenesis of Chloroplasts and Mitochondria, Canberra, 1969.

Received for publication 14 September 1970, and in revised form 28 October 1970.

REFERENCES

1. Armstrong, J. J., S. J. Surzycki, B. Moll, and R. P. Levine. 1971. Genetic transcription and translation specifying chloroplast components in *Chlamydomonas reinhardi*. *Biochemistry.* **10**:692.
2. Arnon, D. 1949. Copper enzymes in isolated chloroplasts. Polyphenol oxidases in *Beta vulgaris. Plant Physiol.* **24**:1.
3. Bennoun, P., and R. P. Levine. 1967. Detecting mutants that have impaired photosynthesis by their increased level of fluorescence. *Plant Physiol.* **42**:1284.
4. Davey, P. J., R. Yu, and A. W. Linnane. 1969. The intracellular site of formation of the mitochondrial protein synthetic system. *Biochem. Biophys. Res. Commun.* **36**:30.
5. Ellis, R. J. 1969. Chloroplast ribosomes: Stereo-specificity of inhibition by chloramphenicol. *Science (Washington).* **158**:477.
6. Ellis, R. J. 1970. Further similarities between chloroplast and bacterial ribosomes. *Planta.* **91**:329.
7. Goodenough, U. W. 1970. Chloroplast division and pyrenoid formation in *Chlamydomonas reinhardi. J. Phycol.* **6**:1.
8. Goodenough, U. W., J. J. Armstrong, and R. P. Levine. 1969. Photosynthetic properties of *ac-31*, a mutant strain of *Chlamydomonas reinhardi* devoid of chloroplast membrane stacking. *Plant Physiol.* **44**:1001.
9. Goodenough, U. W., and R. P. Levine. 1969. Chloroplast ultrastructure in mutant strains of *Chlamydomonas reinhardi* lacking components of the photosynthetic apparatus. *Plant Physiol.* **44**:990.
10. Goodenough, U. W., and R. P. Levine. 1970. Chloroplast structure and function in *ac-20*, a mutant strain of *Chlamydomonas reinhardi*. III. Chloroplast ribosomes and membrane organization. *J. Cell Biol.* **44**:547.
11. Goodenough, U. W., and R. P. Levine. 1971. The effects of inhibitors of RNA and protein synthesis on the recovery of chloroplast ribosomes, membrane organization, and photosynthetic electron transport in the *ac-20* strain of *Chlamydomonas reinhardi. J. Cell Biol.* **50**:50.
12. Goodenough, U. W., and L. A. Staehelin. 1971. Structural differentiation of stacked and unstacked chloroplast membranes: Freeze-etch electron microscopy of wild-type and mutant strains of *Chlamydomonas. J. Cell Biol.* **48**:594.
13. Gorman, D. S., and R. P. Levine. 1965. Cytochrome *f* and plastocyanin: their sequence in the photosynthetic electron transport chain of *Chlamydomonas reinhardi. Proc. Nat. Acad. Sci. U.S.A.* **54**:1665.
14. Hartmann, G., K. O. Honikel, F. Knüsel, and J. Nüesch. 1967. The specific inhibition of the DNA-directed RNA synthesis by rifamycin. *Biochem. Biophys. Acta.* **145**:843.
15. Hoober, J. K., and G. Blobel. 1969. Characterization of the chloroplastic and cytoplasmic

ribosomes of *Chlamydomonas reinhardi*. *J. Mol. Biol.* **41**:121.

16. HOOBER, J. K., P. SIEKEVITZ, and G. E. PALADE. 1969. Formation of chloroplast membranes in *Chlamydomonas reinhardi y-1*: Effects of inhibitors of protein synthesis. *J. Biol. Chem.* **244**:2621.

17. HUDOCK, G. A., G. C. McLEOD, J. MORAVKOVA-KIELY, and R. P. LEVINE. 1964. The relation of oxygen evolution to chlorophyll and protein synthesis in a mutant strain of *Chlamydomonas reinhardi*. *Plant Physiol.* **39**:898.

18. JOHNSON, U. G., and K. R. PORTER. 1968. Fine structure of cell division in *Chlamydomonas reinhardi*. Basal bodies and microtubules. *J. Cell Biol.* **38**:403.

19. KATES, J., and R. F. JONES. 1964. The control of gametic differentiation in liquid cultures of *Chlamydomonas*. *J. Cell. Comp. Physiol.* **63**:157.

20. KUNTZEL, H. 1969. Proteins of mitochondrial and cytoplasmic ribosomes from *Neurospora crassa*. *Nature (London)*. **222**:142.

21. LEVINE, R. P. 1960. Genetic control of photosynthesis in *Chlamydomonas reinhardi*. *Proc. Nat. Acad. Sci. U.S.A.* **46**:972.

22. LEVINE, R. P., and U. W. GOODENOUGH. 1970. The genetics of photosynthesis and of the chloroplast in *Chlamydomonas reinhardi*. *Annu. Rev. Genet.* **4**:397.

23. LEVINE, R. P., and A. PASZEWSKI. 1970. Chloroplast structure and function in *ac-20*, a mutant strain of *Chlamydomonas reinhardi*. II. Photosynthetic electron transport. *J. Cell Biol.* **44**:540.

24. LEVINE, R. P., and D. VOLKMANN. 1961. Mutants with impaired photosynthesis in *Chlamydomonas reinhardi*. *Biochem. Biophys. Res. Commun.* **6**:264.

25. LINNANE, A. W., and P. R. STEWART. 1967. The inhibition of chlorophyll formation in *Euglena* by antibiotics which inhibit bacterial and mitochondrial protein synthesis. *Biochem. Biophys. Res. Commun.* **27**:511.

26. MACKINNEY, G. 1941. Absorption of light by chlorophyll solutions. *J. Biol. Chem.* **140**:315.

27. NEUPERT, W., W. SEBALD, A. J. SCHWAB, P. MASSINGER, and T. BÜCHER. 1969. Incorporation *in vivo* of ^{14}C-labelled amino acids into the proteins of mitochondrial ribosomes from *Neurospora crassa* sensitive to cycloheximide and insensitive to chloramphenicol. *Eur. J. Biochem.* **10**:589.

28. NEUPERT, W., W. SEBALD, A. J. SCHWAB, A. PFALLER, and T. BÜCHER. 1969. Puromycin sensitivity of ribosomal label after incorporation of ^{14}C-labelled amino acids into isolated mitochondria from *Neurospora crassa*. *Eur. J. Biochem.* **10**:585.

29. OHAD, I., P. SIEKEVITZ, and G. E. PALADE. 1967. Biogenesis of chloroplast membranes. I. Plastid dedifferentiation in a dark-grown algal mutant (*Chlamydomonas reinhardi*). *J. Cell Biol.* **35**:521.

30. SIEGAL, M., and H. D. SISLER. 1964. Site of action of cycloheximide in cells of *Saccharomyces pastorianus*. II. The nature of inhibition in a cell-free system. *Biochem. Biophys. Acta.* **87**:83.

31. SMILLIE, R. M., D. GRAHAM, M. R. DWYER, A. GRIEVE, and N. F. TOBIN. 1967. Evidence for the synthesis *in vivo* of proteins of the Calvin cycle and of the electron-transfer pathway on chloroplast ribosomes. *Biochem. Biophys. Res. Commun.* **28**:604.

32. SURZYCKI, S. J. 1969. Genetic functions of the chloroplast of *Chlamydomonas reinhardi*: Effect of rifampin on chloroplast DNA-dependent RNA polymerase. *Proc. Nat. Acad. Sci. U.S.A.* **63**:1327.

33. SURZYCKI, S. J., U. W. GOODENOUGH, R. P. LEVINE, and J. J. ARMSTRONG. 1970. Nuclear and chloroplast control of chloroplast structure and function in *Chlamydomonas reinhardi*. *Symp. Soc. Exp. Biol.* **24**:13.

34. SURZYCKI, S. J., and P. J. HASTINGS. 1968. Control of chloroplast RNA synthesis in *Chlamydomonas reinhardi*. *Nature (London)*. **220**:786.

35. TOGASAKI, R. K., and R. P. LEVINE. 1970. Chloroplast structure and function in *ac-20*, a mutant strain of *Chlamydomonas reinhardi*. I. CO_2 fixation and ribulose-1,5-diphosphate carboxylase synthesis. *J. Cell Biol.* **44**:531.

36. WEHRLI, W., J. NÜESCH, F. KNÜSEL, and M. STAEHELIN. 1968. Actions of rifamycins on RNA polymerase. *Biochem. Biophys. Acta.* **157**:215.

THE EFFECTS OF INHIBITORS OF RNA AND PROTEIN SYNTHESIS ON THE RECOVERY OF CHLOROPLAST RIBOSOMES, MEMBRANE ORGANIZATION, AND PHOTOSYNTHETIC ELECTRON TRANSPORT IN THE ac-20 STRAIN OF CHLAMYDOMONAS REINHARDI

URSULA W. GOODENOUGH and R. P. LEVINE

ABSTRACT

The ac-20 strain of Chlamydomonas reinhardi is characterized by low levels of chloroplast ribosomes when grown mixotrophically. Cells can be transferred to minimal medium and their ribosome levels increase. If, at the time of transfer, cells are exposed to chloramphenicol, an inhibitor of protein synthesis in the chloroplast, or cycloheximide, an inhibitor of protein synthesis in the cytoplasm, ribosome recovery is not affected; however, recovery is blocked by exposure to rifampicin, an inhibitor of chloroplast DNA-dependent RNA polymerase. It is therefore concluded that ac-20 cells suffer from an impaired chloroplast ribosomal RNA synthesis. Mixotrophic ac-20 cells are also characterized by low rates of photosynthetic electron transport, disorganized chloroplast membranes, and a small pyrenoid. If chloramphenicol is applied to transferred cells whose chloroplast ribosome levels have already recovered, recovery of photosynthetic electron transport and of structural integrity does not occur. Under the same conditions, cycloheximide has no effect on recovery. It is concluded that the structural and photosynthetic lesions in ac-20 are a secondary consequence of the low levels of chloroplast ribosomes. Finally, we present evidence that recovery of photosynthetic electron transport requires the transcription of chloroplast DNA. This transcription is apparently triggered by light.

INTRODUCTION

In a previous study of the ac-20 strain of the unicellular green alga, Chlamydomonas reinhardi (6, 12, 18), it was proposed (6) that the primary effect of the ac-20 mutation was to bring about a dramatic reduction in the cells' ability to produce chloroplast ribosomes. It was further proposed that, as a consequence of their greatly reduced levels of chloroplast ribosomes, the ac-20 cells were unable to synthesize certain chloroplast-specific components in normal amounts. These components include several electron-carrier molecules in the photosynthetic electron transport chain (12), the enzyme ribulose-1,5-diphosphate carboxylase (18), and factor(s) necessary for normal chloroplast membrane organization and pyrenoid formation (6).

Our proposal that the ac-20 syndrome results from defective chloroplast protein synthesis was based largely on the sequence of events we observed in so-called transfer experiments. The ac-20 mutation is more strongly expressed if cells are grown mixotrophically (in the light on an

156

acetate-supplemented minimal medium) than if grown phototrophically (in the light on minimal medium). Therefore, when *ac-20* cells are transferred from mixotrophic to phototrophic conditions, a fourfold increase in chloroplast ribosome levels occurs within a few hours. This increase is always followed by, rather than accompanied or preceded by, an increase in levels of the affected chloroplast components. It thus appeared likely that the affected components were dependent on the chloroplast ribosomes for their synthesis.

To further evaluate this interpretation of the *ac-20* syndrome and, in so doing, to further establish the protein-synthesizing capacity of the chloroplast of *C. reinhardi*, experiments have been performed in which antibiotic inhibitors of chloro plastic and cytoplasmic protein synthesis are presented to *ac-20* cells at the time of transfer to minimal medium. The effects of the antibiotics on the recovery of ribosomes and of affected chloroplast components are then studied. The antibiotics used are rifampicin, an inhibitor of DNA-dependent RNA polymerization in the chloroplast (16), chloramphenicol and spectinomycin, two inhibitors of protein synthesis on chloroplast ribosomes (2, 3, 8), and cycloheximide, an inhibitor of cytoplasmic protein synthesis (0, 13).

The results of the experiments reported in this paper indicate that (*a*) the primary lesion in *ac-20* cells is a defective synthesis of chloroplast ribosomal RNA and not a defective synthesis of chloroplast ribosomal protein; (*b*) the recovery of chloroplast membrane organization and photosynthetic electron transport is dependent on protein synthesis on chloroplast ribosomes and is independent of cytoplasmic protein synthesis; (*c*) the recovery of photosynthetic electron transport involves a light-triggered transcription of chloroplast DNA that is distinct from the transcription of information required for chloroplast ribosomal RNA formation. In a separate communication, Togasaki will report results of similar experiments with *ac-20* that monitor the recovery of ribulose-1,5-diphosphate carboxylase in the presence of antibiotics.

MATERIALS AND METHODS

Cells of the *ac-20* strain were grown mixotrophically as described previously (18), except that the concentration of sodium acetate in the culture medium was increased to 0.3% and fresh acetate-supplemented minimal medium (15) was added periodically to the culture as growth proceeded, to assure that the amount of acetate available to the cells did not become limiting.

For transfer experiments, the final volume of the mixotrophic culture was 1–3 liters, depending on how many cells were needed, and the culture had been growing 48–72 hr. Cells were harvested, washed once in minimal medium, and resuspended in minimal medium. All procedures were performed under sterile conditions. Aliquots of 300 ml of the resuspended cells were placed in 500-ml Erlenmeyer flasks. For light-to-dark transfer experiments, the flasks were wrapped with aluminum foil and placed on rotary shakers for 12–16 hr. The foil was then removed, exposing the cells to light for the final stages of the experiment. For light-to-light transfer experiments, the flasks were placed directly in the light. The light intensity was 2000 lux from daylight fluorescent lamps, except in certain experiments with rifampicin to be described below.

For experiments using antibiotics, stock solutions of antibiotics were freshly prepared as described in the previous paper (4); appropriate samples were added at the times indicated, to give the following final concentrations: rifampicin (Mann Research Labs. Inc., New York), 250 µg/ml; chloramphenicol (Sigma Chemical Co., St. Louis, Mo.), 100 µg/ml; spectinomycin (a gift from the Upjohn Co., Kalamazoo, Mich.), 3 µg/ml; cycloheximide (Sigma Chemical Co.), 1 µg/ml.

As pointed out in the previous paper (4), rifampicin is red-orange in color and acts to screen out some of the light available to a culture of cells. In certain experiments with rifampicin, therefore, a culture containing the antibiotic and a control culture surrounded by a bath of a rifampicin solution at the same concentration were both illuminated from below at a light intensity of 8000 lux. Results of such experiments were the same as those carried out at lower light intensities.

At the times indicated in the text, cells were harvested. A portion of the culture was fixed for electron microscopy (6) and the remainder was washed and disrupted by sonication as previously described (11). The resulting chloroplast fragments were used to assay rates of Hill reaction, using 2,6-dichlorophenol-indophenol (DPIP) as an electron acceptor, with a Model 14 Cary recording spectrophotometer (Cary Instruments, Monrovia, Calif.) in the IR mode (7, 14). In some cases, Hill reaction rates were also measured with whole cells, using *p*-benzoquinone as electron acceptor and monitoring light-induced O_2 evolution with a Clarke-type oxygen electrode (Yellow Springs Instrument Co., Yellow Springs, Ohio).

Procedures for counting chloroplast ribosomes from electron micrographs were as described elsewhere (6).

157

Chloroplast Ribosomes

Mixotrophically grown *ac-20* cells (Fig. 1) possess low levels of chloroplast ribosomes (6). When such cells are transferred to minimal medium and placed either in the light (a light-to-light transfer) or in the dark (a light-to-dark transfer), their levels of chloroplast ribosomes increase approximately fourfold (Fig. 2 and reference 6) within a 12 hr period, the rate of increase being more rapid in the light than in the dark (6). A sensitive measure of this increase is provided by the cut-and-weigh technique whereby areas of chloroplast stroma are cut from electron micrographs, the number of chloroplast ribosomes present in the cuttings is counted, the cuttings are weighed, and a value of ribosomes per stroma weight is calculated (6). Typical values for control cells in a transfer experiment are given in Table I.

It is seen in Table I that if rifampicin is given to *ac-20* cells at the time they are transferred to minimal medium, virtually no recovery of chloroplast ribosomes occurs during the ensuing 12 hr. In contrast, if either chloramphenicol or cycloheximide is given at this time, recovery of chloroplast ribosomes is not affected. The values given in Table I are taken from light-to-dark transfer experiments, but comparable results are obtained in light-to-light transfer experiments, as illustrated in Figs. 3–5.

Chloroplast Membrane Organization

Mixotrophically grown *ac-20* cells exhibit three abnormal patterns of chloroplast membrane organization (6): membrane is found in vesicles, in single, unstacked thylakoids (Fig. 1), and in stacked thylakoids where the stack size tends to be larger and the stacks less confluent than in wild-type cells (Fig. 1) (5). The relative proportions of these three types of membrane organization vary considerably from one cell to the next and also from one culture to the next. For example, vesicles may be a prominent feature of cells in one experiment and be only infrequently encountered in cells from another experiment. The basis for this variation is not known, but it necessitates a careful examination of control cultures along with each antibiotic-treated culture before the degree of membrane recovery in a transfer experiment can be assessed.

Following a light-to-light transfer, chloroplast membrane reorganization proceeds after a lag of about 6 hr. By 10–12 hr, membranes are arranged in orderly stacks of two thylakoids; no vesicles and few single thylakoids or high stacks are observed (Fig. 2). If either rifampicin or chloramphenicol is given to the *ac-20* cells at the time of a light-to-light transfer, no recovery of membrane organization occurs during the ensuing 10–12 hr, although the membranes undergo certain rearrangements. In the case of rifampicin-treated cells, the thylakoids tend to lie together (Fig. 3) but they do not stack; in the case of chloramphenicol-treated cells, the membranes associate into poorly defined aggregates (Fig. 4). On the other hand, cycloheximide does not prevent the assumption of an orderly stacking pattern (Fig. 5).

When *ac-20* cells are subjected to a light-to-dark transfer, a dramatic transformation in chloroplast membrane organization takes place during the ensuing 12–16 hr in the dark. Most striking is the accumulation of huge aggregates of membrane that we have termed membrane "piles" (6). In the present series of experiments we have found that such piles accumulate even when rifampicin, cycloheximide, or chloramphenicol is present; in other words, pile formation is not prevented by any of the inhibitors of protein synthesis that we have used.

After the 12–16 hr dark incubation that follows a light-to-dark transfer, cells are routinely returned to the light. When control cells are exposed to light, the membrane piles rapidly disperse and appear to give rise to smaller stacks that arrange themselves in the orderly, two-thylakoid arrays found in "recovered" cells (6). The membrane piles also disappear in the antibiotic-treated cells after exposure to light, but the course of events that follows their dispersal depends on the antibiotic and on the time the antibiotic is given to the cells.

If either rifampicin or chloramphenicol is given to cells at the start of a light-to-dark transfer experiment and if the cells are placed in the light at the end of 12–16 hr, the membrane does not come to assume an orderly organization as in the control. Instead, the membrane piles disperse and the membrane returns to its unstacked, disordered, mixotrophic state. If, on the other hand, the cells are transferred to minimal medium and allowed to incubate in the dark for 12–16 hr *before* either rifampicin or chloramphenicol is added, the antibiotics have no effect on the ensuing light-stimulated membrane reorganization; after several hours in the light the treated cells are indistinguishable from the controls.

Cyclcheximide has quite a different effect. Regardless of when it is administered, it is without effect on chloroplast membrane organization: in its presence, membrane piles accumulate in the dark and give rise, in the light, to orderly arrays of small stacks.

Pyrenoid Formation

We previously reported that a pyrenoid is either lacking or rudimentary in mixotrophic ac-20 cells (6). Further examination of this point indicates that in fact most mixotrophic ac-20 cells probably possess rudimentary pyrenoids, but because of their small size, they are only rarely encountered in section. If 100 cells of comparable size are scored for the presence or absence of pyrenoid material in thin section, 40 are found to exhibit pyrenoids when the cells are from a wild-type culture, whereas seven exhibit pyrenoids when the cells are from a mixotrophic ac-20 culture.

This type of counting procedure is meaningful only if cells are the same size: large pyrenoids in small cells will, for example, be encountered more frequently than large pyrenoids in large cells, and so on. Since ac-20 cells undergo a cell division during the course of a transfer experiment (6), it is not possible to give a quantitative estimate of pyrenoid frequencies as they are observed during a transfer experiment. Qualitatively, however, it is quite apparent that pyrenoids undergo a dramatic increase in size during a transfer experiment, that both chloramphenicol and rifampicin prevent this increase, and that cycloheximide does not affect the increase. As reported previously (6), pyrenoid recovery proceeds during the dark period of a light-to-dark transfer experiment.

Hill Activity

The Hill reaction measures the light-induced reduction of an electron acceptor where water is the electron donor. In this paper, Hill reaction rates are given for the photoreduction of DPIP by chloroplast fragments. These experiments were also performed with whole cells using p-benzoquinone as the electron acceptor, and comparable results were obtained. Thus the effects of the inhibitors that are reported below cannot be attributed to antibiotic-produced differences in the way chloroplast membranes fragment when subjected to sonic disintegration.

As reported by Levine and Paszewski (12), mixotrophic ac-20 cells have low rates of Hill activity, commonly one-fourth the rates of wild-type cells. In Figs. 6–9 of the present paper it is seen that the rates of mixotrophic ac-20 cells vary between 45 and 85 μmoles of DPIP reduced/hr per mg chlorophyll (chl). A comparable range of variation in rates is observed in wild-type cells from one experiment to the next. Following transfer to minimal medium, Hill reaction rates in ac-20 cells increase, after a lag, in the light; in the dark, the rates do not increase, but after light is provided, recovery proceeds without a lag (12). In the present experiments, only initial and final rates were measured for a given set of experimental conditions. Therefore, the straight lines in Figs. 6–9 do not represent the actual kinetics of recovery.

Fig. 6 shows that if rifampicin is given to mixotrophic ac-20 cells at the start of a light-to-light transfer experiment, recovery of Hill activity is inhibited although not, in this experiment, inhibited completely. This experiment was carried out at high light intensities in order to make certain that the red color of the rifampicin was not screening out light and thereby preventing recovery. The high light intensities may, however, produce a partial breakdown of the inhibitor. Other difficulties in working with rifampicin are discussed in the preceding paper (4). It is also seen in Fig. 6 that if rifampicin is added at the end of the dark incubation period that follows a light-to-dark transfer, a complete inhibition of recovery of Hill activity is effected.

When chloramphenicol is given to cells at either the onset of a light-to-light transfer experiment or before providing light in a light-to-dark transfer, no recovery of Hill activity occurs (Fig. 7). There is some indication that chloramphenicol may be inhibitory to the photosynthetic process itself in short-term experiments (1), and we therefore repeated these experiments with a second inhibitor of chloroplast protein synthesis, spectinomycin. As seen in Fig. 8, identical results were obtained.

Fig. 9 shows that cycloheximide is without effect on the recovery of Hill activity, whether given at the beginning of a light-to-light transfer or at the end of a light-to-dark incubation.

159

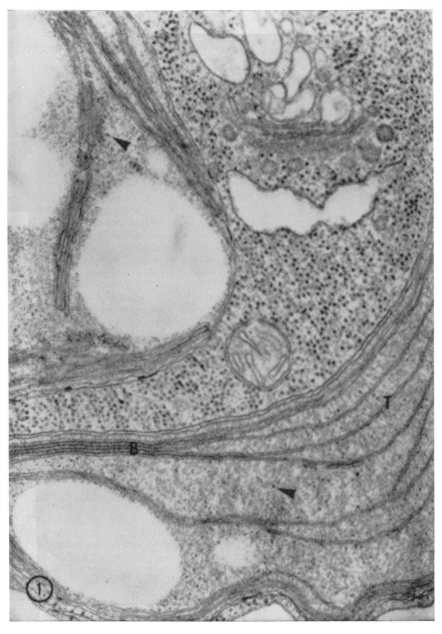

FIGURE 1   Portion of an *ac-20* cell grown mixotrophically. Chloroplast ribosomes (arrowheads) are sparse. Thylakoids (T) are unstacked except for a wide band (B) beneath the chloroplast envelope. × 67,000.

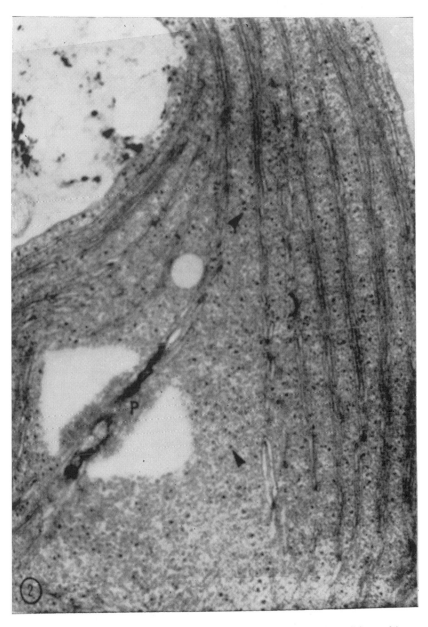

FIGURE 2 Portion of an *ac-20* cell transferred from mixotrophic to phototrophic conditions and incubated in the light for 8 hr. Chloroplast ribosomes (arrowheads) have recovered (compare with Fig. 1). Thylakoids are in the process of becoming ordered into the concentric arrays of two-thylakoid stacks that typify phototrophic *ac-20* cells (6). A grazing section of the pyrenoid (*P*) is included. × 72,000.

TABLE I

Effect of Antibiotics on the Recovery of Chloroplast Ribosomes in ac-20 Cells during Light-to-Dark Transfer Experiments, as Determined by the Cut-and-Weigh Technique

Antibiotic added	Initial chloroplast ribosome level (ribosomes stroma weight)	Final chloroplast ribosome level 16 hr after transfer	% inhibition by antibiotic
Control	20	110	—
Rifampicin (250 µg/ml)	20	30	89
Chloramphenicol (100 µg/ml)	20	106	4
Cycloheximide (1 µg/ml)	20	114	0

DISCUSSION

The experiments reported in this paper indicate that the low levels of chloroplast ribosomes found in mixotrophic ac-20 cells are the consequence of defective chloroplast DNA-dependent RNA synthesis. Surzycki has shown (16) that the 5S, 16S, and 23S species of chloroplast ribosomal RNA (rRNA) are among the products of chloroplast DNA-dependent RNA synthesis and that these species of rRNA are greatly reduced in amount in mixotrophic ac-20 (6, 17); it therefore seems most reasonable to conclude that a primary, if not the primary, lesion suffered by mixotrophic ac-20 cells is a reduced ability to synthesize chloroplast rRNA. This lesion is apparently not accompanied by a reduction in the synthesis of chloroplast ribosomal proteins or in the synthesis of the chloroplast DNA-dependent RNA polymerase enzyme. A sizable store of these proteins must exist in mixotrophic ac-20 cells, for when chloroplast rRNA synthesis is stimulated by transfer to minimal medium, chloroplast ribosomes assemble in the presence of either chloramphenicol or cycloheximide. In other words, in order for chloroplast ribosome levels to recover, RNA synthesis in the chloroplast must occur but protein synthesis in the chloroplast or in the cytoplasm need not occur.

Our experiments also indicate that three other aspects of the ac-20 "syndrome"—disorganized membrane structure, rudimentary pyrenoids, and defective Hill activity—are a direct consequence of the reduced levels of protein synthesis in the mixotrophic ac-20 chloroplast. In the presence of chloramphenicol, recovery of these three parameters is not observed even after chloroplast ribosome levels have recovered; in contrast, cycloheximide does not prevent their recovery. Thus we can conclude that translational steps on the chloroplast ribosomes of C. reinhardi are themselves sufficient to bring about dramatic changes in chloroplast membranes, in pyrenoid formation, and in photosynthetic capacity. These results are in full accord with other studies of C. reinhardi (1, 4, 6, 9, 12, 18) and provide further support for the conclusion that chloroplast protein synthesis plays a significant role in the construction of a functional chloroplast in this organism.

The application of protein-synthesis inhibitors to ac-20 cells at various times after transfer to minimal medium has provided a greater understanding of the recovery process itself. First, it has become apparent that the dramatic piling up of thylakoid membranes that occurs during the dark period of a light-to-dark transfer is not an obligate phase of recovery. Rather, it appears to be a response of the ac-20 thylakoid membranes to a dark incubation in minimal medium. The response does not require RNA or protein synthesis since it occurs in the presence of rifampicin, chloramphenicol, and cycloheximide. Moreover, if cells are presented with either rifampicin or chloramphenicol (but not cycloheximide) at the start of a light-to-dark transfer experiment and are exposed to light after 12 hr, the large piles disperse but they do not give rise to normal membrane structures. Thus the membrane piles do not represent, as was previously suggested (6), stores of membrane material that are "potentiated" for the formation of normally organized thylakoids once light is provided.

In spite of the fact that the visible alterations in membrane conformation that occur during the dark period are not morphological manifestations of the recovery process, it is nonetheless clear that alterations in ac-20 chloroplast membranes that depend upon chloroplast protein synthesis do occur during the dark period. This is demonstrated by the observation that if either rifampicin or chloramphenicol (but not cycloheximide) is given at the *start* of the dark period, normal membrane configurations never form in the ensuing light period, whereas if either rifampicin, chloramphenicol, or cycloheximide is given to cells at the *end* of the dark period, membrane reorganization proceeds normally when the cells are exposed to light. Thus the process of membrane reorgani-

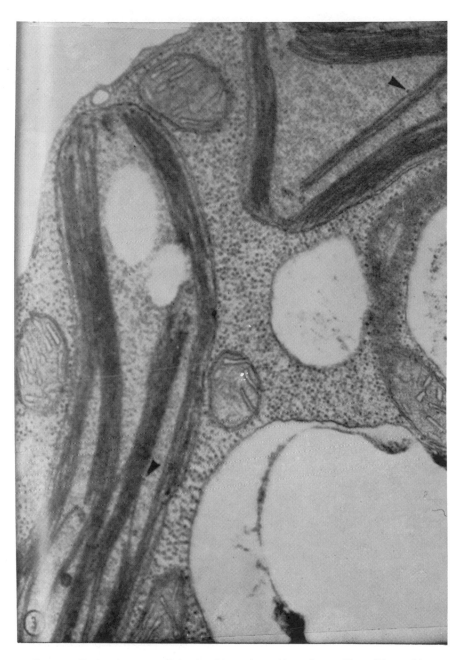

FIGURE 3 Portion of an *ac-20* cell transferred from mixotrophic to phototrophic conditions and incubated in the light in the presence of 250 μg/ml rifampicin for 11 hr. No recovery of chloroplast ribosomes (arrowheads) has occurred (compare with Figs. 1 and 2). Thylakoids associate in large bundles but stacking does not usually occur between these thylakoids. × 71,000.

163

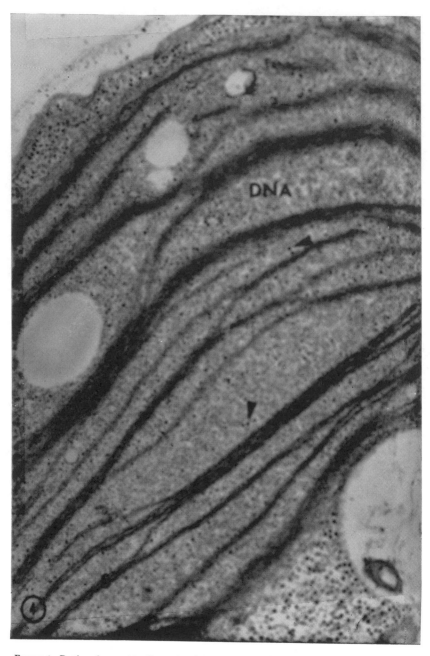

FIGURE 4   Portion of an *ac-20* cell transferred from mixotrophic to phototrophic conditions and incubated in the light in the presence of 100 μg/ml chloramphenicol for 10 hr. Chloroplast ribosomes (arrowheads) have recovered (compare with Figs. 1 and 2). A short region of normal stacking (S) is included in the field, but most of the membrane lies in ill-defined aggregates (the poor definition of these aggregates is not the result of an oblique sectioning angle; most cells in the sample exhibit a preponderance of similar profiles). A region of chloroplast DNA is indicated (*DNA*). × 71,000.

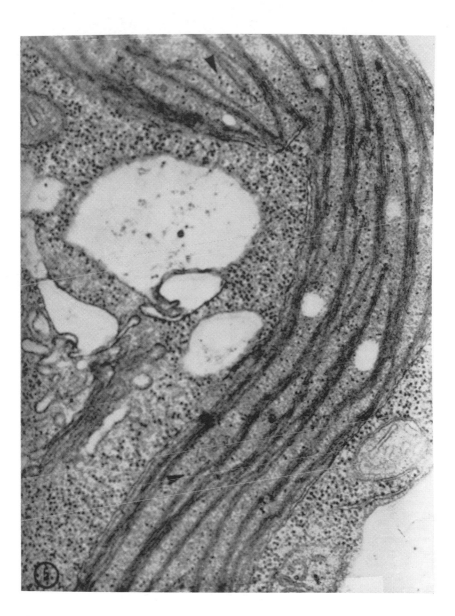

FIGURE 5 Portion of an *ac-20* cell transferred from mixotrophic to phototrophic conditions and incubated in the light in the presence of 1 µg/ml cycloheximide for 12 hr. Chloroplast ribosomes (arrowheads) have recovered (compare with Figs. 1 and 2). Thylakoids are in the process of becoming ordered into two-thylakoid stacks. × 71,000.

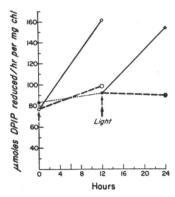

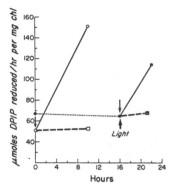

FIGURE 6 The effect of 250 μg/ml rifampicin on the recovery of DPIP-Hill activity in mixotrophic *ac-20* cells transferred to minimal medium at 0 hr. Open symbols represent values from a light-to-light transfer experiment, and closed symbols represent values from a light-to-dark transfer experiment. Control values are indicated by circles and solid lines. Values from rifampicin-treated cells are symbolized by hexagons and dashed lines. For the light-to-dark transfer experiment, the reaction rates at the beginning and at the end of the dark incubation period are given; the two values are connected by a dotted line. The time that light is provided in the light-to-dark transfer experiment is indicated. Rifampicin was added to the experimental cultures at the times indicated by the thinner arrows.

FIGURE 7 The effect of 100 μg/ml chloramphenicol on the recovery of DPIP-Hill activity in mixotrophic *ac-20* cells transferred to minimal medium at 0 hr. Symbols are as described for Fig. 6 except that values from chloramphenicol-treated cells are symbolized by squares, and the thinner arrows indicate the times of addition of chloramphenicol.

zation in *ac-20* appears to involve two phases: a light-independent, synthetic phase that is dependent on protein synthesis in the chloroplast but not in the cytoplasm, and a light-dependent, organizational phase that is independent of protein synthesis, during which the "potentiated" membranes formed in the dark are rapidly arranged into their more orderly morphological configurations. The existence of these two phases cannot be readily detected in light-to-light experiments because, in such experiments, synthesis and organization occur concurrently.

The stimulation of membrane reorganization by light apparently involves an interaction between light and chloroplast membranes that is not mediated by RNA or protein synthesis. Our experiments also indicate a second role for light in the recovery process. If cells are exposed to rifampicin, chloramphenicol, or spectinomycin (but not cycloheximide) at the end of the dark period (a time, it will be recalled, when chloro-

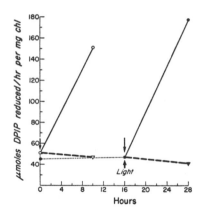

FIGURE 8 The effect of 3 μg/ml spectinomycin on the recovery of DPIP-Hill activity in mixotrophic *ac-20* cells transferred to minimal medium at 0 hr. Symbols are as described for Fig. 6 except that values from spectinomycin-treated cells are symbolized by triangles, and the thinner arrows indicate the times of addition of spectinomycin.

plast ribosomes have already formed), no recovery of Hill activity occurs, even though membrane reorganization is not blocked. Thus light is apparently required to trigger, and perhaps also to sustain, a chloroplast-located protein synthesis

166

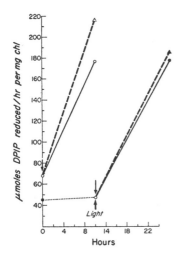

FIGURE 9   The effect of 1 μg/ml cycloheximide on the recovery of DPIP-Hill activity in mixotrophic *ac-20* cells transferred to minimal medium at 0 hr. Symbols are as described for Fig. 6 except that values from cycloheximide-treated cells are symbolized by triangles, and the thinner arrows indicate the times of addition of cycloheximide.

that produces a photosynthetic apparatus capable of greatly stimulated Hill activity. The fact that rifampicin produces a complete block in recovery of Hill activity under these experimental conditions strongly suggests that light is first required to stimulate a DNA-dependent RNA synthesis within the chloroplast, one that cannot or does not occur in the dark. The RNA so produced is apparently then translated on chloroplast ribosomes (unless these are blocked by chloramphenicol or spectinomycin). Whether light is also required for the translational phase of the process cannot be determined from these experiments. We believe, then, that our experiments indicate the existence of a light-stimulated messenger RNA synthesis from chloroplast DNA in *C. reinhardi*.

In summary, it appears that the *ac-20* mutation produces a primary defect in chloroplast rRNA synthesis; it may produce a more generalized inhibition of all chloroplast DNA-directed RNA synthesis, but our experiments were not designed to examine this possibility. As a consequence of this defective rRNA synthesis, chloroplast ribosomes are not produced at normal levels and normal synthesis of proteins required for the Hill

reaction, for membrane organization, and for pyrenoid formation cannot occur. Once cells have been transferred to minimal medium and chloroplast ribosome levels have increased, chloroplast protein synthesis also increases. The synthesis of component(s) required for normal membrane organization and pyrenoid formation can apparently occur in either the light or the dark, but recovery of Hill activity appears to require a light-stimulated transcription of chloroplast DNA and thus occurs only in the light. Cytoplasmic protein synthesis appears not to be involved in any of the aspects of the *ac-20* recovery process that we have studied; indeed, the absence of any effect of cycloheximide is most dramatic, for in short-term synchronous-culture experiments with *C. reinhardi*, cycloheximide produces a rapid and complete inhibition of most synthetic processes (1).

The existence of a Mendelian gene that exerts an apparent control over chloroplast DNA transcription is potentially of considerable interest, as we have discussed elsewhere (6, 10). However, the strong influence of mixotrophic *versus* phototrophic growth conditions over the expression of the *ac-20* mutation leads us to question how direct this "control" really is.

Initial experiments on the effects of chloramphenicol and cycloheximide on recovery of Hill activity were performed in this laboratory by Dr. A. Paszewski. We acknowledge the excellent technical assistance of Miss Diane Dwyer and the continued interest in this research given by Dr. S. J. Surzycki and Miss J. J. Armstrong.

Research was supported by a postdoctoral fellowship to Dr. Goodenough from the National Institutes of Health (GM 24306), by the Maria Moors Cabot Foundation for Botanical Research, Harvard University, and by Research Grant GB 1866 from the National Science Foundation.

An abbreviated report of this research was presented at the Symposium on the Biogenesis of Chloroplasts and Mitochondria, Canberra, 1969.

Received for publication 14 September 1970, and in revised form 28 October 1970.

REFERENCES

1. ARMSTRONG, J. J., S. J. SURZYCKI, B. MOLL, and R. P. LEVINE. 1971. Genetic transcription and translation specifying chloroplast components in *Chlamydomonas reinhardi. Biochemistry.* **10**: 692.

2. ELLIS, R. J. 1969. Chloroplast ribosomes: Stereo-

specificity of inhibition by chloramphenicol. *Science (Washington).* **158:**477.

3. ELLIS, R. J. 1970. Further similarities between chloroplast and bacterial ribosomes. *Planta.* **91:**329.

4. GOODENOUGH, U. W. 1970. The effects of inhibitors of RNA and protein synthesis on chloroplast structure and function in wild-type *Chlamydomonas reinhardi. J. Cell Biol.* **50:** 35.

5. GOODENOUGH, U. W., and R. P. LEVINE. 1969. Chloroplast ultrastructure in mutant strains of *Chlamydomonas reinhardi* lacking components of the photosynthetic apparatus. *Plant Physiol.* **44:** 990.

6. GOODENOUGH, U. W., and R. P. LEVINE. 1970. Chloroplast structure and function in *ac-20*, a mutant strain of *Chlamydomonas reinhardi.* III. Chloroplast ribosomes and membrane organization. *J. Cell Biol.* **44:**547.

7. GORMAN, D. S., and R. P. LEVINE. 1966. Photosynthetic electron transport chain of *Chlamydomonas reinhardi.* VI. Electron transport in mutant strains lacking either cytochrome 553 or plastocyanin. *Plant Physiol.* **41:**1648.

8. HOOBER, J. K., and G. Blobel. 1969. Characterization of the chloroplastic and cytoplasmic ribosomes of *Chlamydomonas reinhardi. J. Mol. Biol.* **41:**121.

9. HOOBER, J. K., P. SIEKEVITZ, and G. E. PALADE. 1969. Formation of chloroplast membranes in *Chlamydomonas reinhardi y-1*: Effects of inhibitors of protein synthesis. *J. Biol. Chem.* **244:**2621.

10. LEVINE, R. P., and U. W. GOODENOUGH. 1970 The genetics of photosynthesis and of the chloroplast in *Chlamydomonas reinhardi. Annu. Rev. Genet.* **4:** 397.

11. LEVINE, R. P., and D. S. GORMAN. 1966. Photosynthetic electron transport chain of *Chlamydomonas reinhardi.* III. Light-induced absorbance changes in chloroplast fragments of the wild type and mutant strains. *Plant Physiol.* **41:** 1293.

12. LEVINE, R. P., and A. PASZEWSKI. 1970. Chloroplast structure and function in *ac-20*, a mutant strain of *Chlamydomonas reinhardi.* II. Photosynthetic electron transport. *J. Cell Biol.* **44:** 540.

13. SIEGAL, M., and H. D. SISLER. 1964. Site of action of cycloheximide in cells of *Saccharomyces pastorianus.* II. The nature of inhibition in a cell-free system. *Biochim. Biophys. Acta.* **87:**83.

14. SMILLIE, R. M. 1962. Photosynthetic and respiratory activities of growing pea leaves. *Plant Physiol.* **37:**716.

15. SUEOKA, N. 1960. Mitotic replication of deoxyribonucleic acid in *Chlamydomonas reinhardi. Proc. Nat. Acad. Sci. U.S.A.* **46:**83.

16. SURZYCKI, S. J. 1969. Genetic functions of the chloroplast of *Chlamydomonas reinhardi:* Effect of rifampin on chloroplast DNA-dependent RNA polymerase. *Proc. Nat. Acad. Sci. U.S.A.* **63:** 1327.

17. SURZYCKI, S., and P. J. HASTINGS. 1968. Control of chloroplast RNA synthesis in *Chlamydomonas reinhardi. Nature (London).* **220:**786.

18. TOGASAKI, R. K., and R. P. LEVINE. 1970. Chloroplast structure and function in *ac-20*, a mutant strain of *Chlamydomonas reinhardi.* I. CO_2 fixation and ribulose-1,5-diphosphate carboxylase synthesis. *J. Cell Biol.* **44:**531.

AUTHOR INDEX

KEY-WORD TITLE INDEX

Guide to Current Research

The research summaries appearing in the following section were obtained through a search of the Smithsonian Science Information Exchange data base conducted in July, 1973.

The Exchange annually registers 85,000 to 100,000 notices of current research projects covering a wide range of disciplines and sources of support. SSIE endeavors to retain up to two full years of current research information in its active file. The selection of summaries appearing in this section does not represent the complete SSIE collection of information on this topic, but, rather, has been specifically tailored to reflect the scientific content of this particular volume. A limited number of summaries may have been omitted because clearance for publication by the supporting agency or organization was not received prior to the publication date.

SSIE is the only, single source for information on ongoing and recently terminated research in all areas of the life, physical, behavioral, social and engineering sciences. The SSIE file is updated daily by a professional staff of scientists utilizing a comprehensive and flexible system of hierarchical indexing. Retrieval of subject information is conducted by these same specialists, using computer-connected, video display terminals which allow instant access to the entire data base and on-line refinement of search strategies. SSIE offers an information service unequalled anywhere: comprehensive and vital information on who is conducting what research where and under whose support.

More current information, and in some cases expanded coverage, on the topic considered in this volume is available directly from SSIE. This information is offered at modest cost in the form of custom searches of the SSIE file designed specifically to meet the user's need or as an update of the subject search in this section. For more information on SSIE, contact MSS or write directly to the Smithsonian Science Information Exchange, 1730 M Street, N.W., Washington, D.C. 20036. Subject search or updated package requirements may be discussed with SSIE scientists by calling the Exchange at (202) 381-5511.

171

COMBINED EFFECT OF RADIATION AND
ENVIRONMENTAL CONTAMINANTS ON DNA REPAIR
MECHANISM,
H. ALTMANN, Austrian Atom. En. Study Group,
Vienna, Austria

Sensitivity of Chlorella cells to
mercury ions and the uptake of Hg ions in
labelled form are investigated. DNA-repair
replication as well as semiconservative DNA
synthesis will be tested after incubation of
cells (lymphocytes, E.coli and Chlorella)
with different chemical compounds. Special
emphasis will be given to pesticides, drugs,
metal ions and other environmental
contaminants.

BIOGENESIS OF CHLOROPLAST MEMBRANES OF
CHLAMYDOMONAS REINHARDI,
D.P. BECK, Harvard University, School of
Arts, Cambridge, Massachusetts 02138

The relative times of synthesis of
specific chloroplast membrane proteins and
lipids are being investigated in
synchronously growing Chlamydomonas
reinhardi. Cells of C. reinhardi can be
induced to divide synchronously in response
to a light-dark cycle. Under this
condition, many cellular constituents are
known to increase at specific times during
the growth cycle. In these studies, a
double-labelling technique is employed to
determine when specific proteins and lipids
are synthesized and incorporated into the
membranes of the chloroplast. Proteins are
separated by disc gel electrophoresis in
SDS-urea gels and the radiometric data are
analyzed statistically by computer.
Membrane lipids are separated by thin layer
chromatography and measured both

radiometrically and chemically. Studies are
planned which will determine the effects of
antibiotics on the lipid and protein
synthesis and incorporation into membranes.
The aim of the program described above is to
elucidate the relationship between membrane
lipid and protein synthesis, and it should
answer some questions concerning the
mechanism of chloroplast membrane
biogenesis.

MECHANISM OF OXYGEN PRODUCTION IN
PHOTOSYNTHESIS,
N.I. BISHOP, Oregon State University,
Agricultural Experiment Sta., Corvallis,
Oregon 97331

OBJECTIVE: Determine, if possible, the
mechanism whereby green plants produce
oxygen. The problem is currently under
study from two angles; the examination of
mutations of the green algae Scenedesmus in
which it is known that the mechanism of
oxygen production has been hindered
genetically and the biochemical analysis on
the various reactions in which a naturally
occurring quinone, Plastoquinone, is known
to be required. This quinone has been shown
to function in reactions leading to oxygen
evolution.
APPROACH: NOT-PROVIDED
PROGRESS: Research activities included
studies on the characteristics of pigment
mutants of Scenedesmus blocked at specific
points in carotenoid biosynthesis, analyses
of photosynthetic mutants blocked at
cytochrome f-552, improved methods of mutant
enrichment and isolation of specific mutants
of photsystem II, and preliminary
observations on certain antibiotics which
alter transcription and translational
properties of chloroplast DNA and RNA.

Detailed studies on the five types of
carotenoid mutants mentioned in last year's
report have shown that the capacity for
oxygen evolution is strongly dependent upon
the formation of the carotenes, and perhaps
the xanthophylls. Also the formation of the
characteristic double-stacking of the
chloroplast lamellae occurs only in those
mutants which form normal carotenoids. The
fundamental role of the carotenoids appears
to be in the development of a normal
chloroplast membrane which determines
photosystem II activity. Photosystem I
activity is not affected even though gross
alteration of membrane structure has
occurred. Increased recovery of mutants has
been made possible by employing an organic
dye plus light and oxygen to kill
selectively normal algal cells from cultures
treated with various mutagens. Additional
methods allow the selection of mutants
blocked on the oxidizing side of photosystem
II rather than on the reducing side.
Preliminary studies of the action of
chloramphenicol, cycloheximide, rifamycin,
spectinomycin, and streptomycin are in
progress.

HEREDITARY CONTROL OF ORGANELLE STRUCTURE
AND FUNCTION,
J.E. BOYNTON, Duke University, Graduate
School, Durham, North Carolina 27706

Our objective is to elucidate the
hereditary control of organelle structure
and function by studying Mendelian (nuclear)
and non-Mendelian (presumably organelle)
mutations affecting chloroplast and
mitochondrial functions. The green alga,
Chlamydomonas reinhardti, is a model system
which allows isolation and characterization
of mutations lacking specific chloroplast

functions (acetate requirers) and mutations
with altered mitochondrial functions
(obligate photoautotrophs). In no other
species can mutations of the chloroplast and
mitochondrion be investigated simultaneously
to determine the genetic interrelationship
of both organelles to the nucleus and to
each other. One well- characterized
non-Mendelian genetic system exhibiting
uniparental (UP) inheritance is well known
in C. reinhardti. Quite possibly UP genes
reside in the chloroplast DNA. Recently we
have identified and begun characterization
of a second non-Mendelian genetic system
which may be mitochondrial.

GENETIC ANALYSIS OF THE BIOLOGICAL CLOCK,
V.G. BRUCE, Princeton University, Graduate
School, Princeton, New Jersey 08540

The basic objective of this research is
to develop the technique to undertake a
genetic analysis of the biological clock in
Chlamydomonas. Little is known about the
mechanism of the biological clock at the
cellular or at the biochemical level and it
seems likely that a genetic analysis could
contribute significantly to our
understanding of this problem. Recent work
done by Dr. Bruce demonstrates the
feasibility of conducting such a genetic
analysis. Clock mutants have been found and
several genetic crosses have been carried
out. However, the practical exploitation of
the system requires greatly increased
capacity to assay the rhythms of individual
cultures than is currently available. Dr.
Bruce will develop the required
instrumentation to provide this capability,
as well as carry out the necessary genetic
experiments to determine the basic nature of
the genetic organization of the system.

MEIOSIS AND REPRODUCTION,
K. CHIANG, Univ. of Chicago, Graduate School,
Chicago, Illinois 60637

Meiosis occurs only as an integral part
of the reproductive process in human and
other higher organisms. The basic
mechanism(s) underlying the induction,
regulation and genetic recombination of
meiosis in sexually reproducing species
remains scarcely explored and poorly
understood.
Using Chlamydomonas reinhardti as a
model organism, we propose to investigate
the molecular mechanisms of meiosis. The
unique characteristics of meiosis, i.e.,
synapsis, crossing-over and reduction in
chromosome number, will be studied by
physical, chemical and genetic techniques.
The regulation and control of the induction
of this meiotic division will also be
investigated. Subsequent to these basic
studies we intend to explore the possibility
of the experimental regulation of the
meiotic process by artificial means; we will
test whether the occurrence of meiotic cell
division itself and the frequency of
synaptic crossing-over in meiosis can be
controlled.

MOLECULAR MECHANISM OF NON-MENDELIAN
INHERITANCE,
K. CHIANG, Univ. of Chicago, School of
Medicine, Chicago, Illinois 60637

This award involves continuation of the
support provided under grant GB-19338 for
Dr. Chiang's research program of the
molecular mechanism of non-Mendelian
inheritance using Chlamydomonas as a model
system. In this system only the mt plus

(plus mating type) parents, not the mt minus (minus mating type) parents, transmit non-Mendelian genes into all the meiotic progeny. Experiments are underway to determine the capacity for continuous replication and to detect any possible modification in the transcription of the mt minus chloroplast or other cytoplasmic DNAs after zygote formation. The molecular details of physical recombination between the mt plus and mt minus chloroplast DNAs during meiosis are also being analyzed. Attempts are also being made to determine whether the mt minus gametic pyrenoid consistently disintegrates during the structural reorganization of two parental chloroplasts after their fusion, and to ascertain whether this situation can be altered by conditions which will induce bi-parental inheritance of non-Mendelian genes. Other problems being investigated include: 1) whether the pattern of uniparental transmission of the non-Mendelian determinants is controlled by the nuclear mating type gene or is regulated autonomously by the non-Mendelian genetic determinants; 2) isolation of antibiotic-resistant uniparental mutants and characterization of cytoplasmic DNA and ribosomal RNA in such mutants.

THE KINETICS OF HYDROGEN EVOLUTION INDUCED BY LIGHT IN SOME GREEN ALGAE, H. GAFFRON, Florida State University, Inst. of Molecular Biophysics, Tallahassee, Florida 32306

The work reported last year has been and will be continued because of a number of

new, unexpected and interesting results.
For instance, photochemical H2 evolution of
the kind described in earlier publications
is completely separated from CO2 evolution.
The latter precedes and accompanies the
photodissimilation as a typical enzymatic
dark reaction. The rate of the light-induced
H2 reaction can be doubled or tripled when
the better known normal photosynthetic
reactions such as carbon dioxide reduction
and photophosphorylation are removed by
careful heating or through poisoning. Even
then the rate of H2 evolution measured over
a period of hours is much lower than
expected, considering that it ought to be
the expression of the basic consequence of
pigment excitation. This puzzle was solved
when it was found that there are two steps
in the kinetics of hydrogen evolution. An
initial very fast phase which was light
dependent and could not be easily light
saturated, and a second phase, the rate of
which depended entirely on the slow
production of endogenous hydrogen donors.
We assume therefore that the first phase is
fed from a small pool of accumulated
hydrogen donors. This pool is directly
connected with the electron transport chain
and the speed of its light-induced
consumption reflects the true capacity of
the pigment hydrogenase combination.

PHOTOPRODUCTION OF MOLECULAR HYDROGEN IN
GREEN ALGAE AND PURPLE BACTERIA,
H. GAFFRON, Florida State University, School
of Arts, Tallahassee, Florida 32306

     Research in photobiology, specifically
photosynthesis and in particular the
evolution of molecular hydrogen promoted by
light in hydrogenase - containing green and
blue - green algae as well as purple

bacteria will be continued.

Experiments now in progress center on an unexpected problem, namely the conspicious difference in the mechanism for hydrogen photoproduction in green algae and in photosynthetic bacteria. The bacteria have to rely on photophosphorylation for the release of molecular hydrogen, while the hydrogenase-containing algae give the highest rate of hydrogen evolution when phosphorylations are abolished with the aid of specific inhibitors. Mass spectograph investigations in flashing light, in mixed illumination (Emerson effect) and in direct combination wth observations of fluorescence will continue.

BIOPHYSICS OF PLANT GROWTH REGULATION, P.B. GREEN, Stanford University, School of Humanities, Palo Alto, California 94305

By following, with high spatial and temporal resolution, the course of rate response to modulation of turgor pressure, it is hoped to identify the major components of of the growth process in higher plants. This technique has revealed a governor-like control of rate in the alga Nitella. After difficult corrections for elastic phenomena, a similar governor appears detectable in Avena coleoptiles. Current efforts are, in part, directed at finding how auxin changes the balance setting of the governor. The major goal, however, is finding the correct analytical expression, preferably comprised of terms measurable in vivo, which is truly comprehensive in that the action of any agent influencing growth can be identified with one or more components. The growth records obtained so far are complex but reproducible. They provisionally indicate that auxin changes two parameters which

control the yield threshold of the wall. The
resultant movement of the threshold to lower
turgor appears for the stimulation in rate
brought on by auxin.

CONTROL OF DNA METABOLISM IN CHLAMYDOMONAS,
S.H. HOWELL, Univ. of California, Graduate
School, San Diego, California 92038

The DNA metabolic processes in the
synchronous mitotic and meiotic division
cycles of Chlamydomonas reinhardi are being
studied by isolating conditional mutants
with lesions in mitotic function. A subset
of mutants is being selected which affect
mitotic DNA metabolic processes, such as DNA
synthesis and UV repair. These mutants are
being screened for the amount of DNA
synthesis, DNA breakage, and rejoining
enzyme activities. The activity of the
characterized conditional mutants is being
tested under restrictive conditions of known
cyclic DNA metabolic functions, such as DNA
synthesis, in synchronous mitosis and
meiosis. Special attention is being given to
effects of these conditional mutants on
genetic recombination and the formation of
bi- parental recombinant molecules.

GENETIC CONTROL OF DEVELOPMENT IN VOLVOX,
R.J. HUSKEY, Univ. of Virginia, School of
Arts, Charlottesville, Virginia 22903

The precise control of early
development and cellular differentiation has
been difficult to analyze because of the
sensitivity of these early stages in most

experimental organisms. Research is underway to begin detailed genetic analysis of the development of a easily manipulated organism, the colonial alga Volvox, which undergoes early stages of development similar to those of most higher animals. Because the cleavage, inversion and differentiation steps occur in aqueous medium in nature, they have been easily studied in axenic culture. Using recently perfected techniques chemical mutagenesis is being used for the first time to induce stable genetic variants of Volvox. Selective techniques are being applied to isolate temperature sensitive mutations in morphogenetic genes as well as mutations in other genes. Mutant strains are being used to determine the sequence and duration of gene action during development. Mutant gene expression is being observed in the electron microscope and described on the biochemical level. These results are being correlated with gross developmental changes. In addition to other advantages, Volvox also has an easily controllable induction system which regulates a major developmental response pattern, the formation of sexual gametes. The selection and isolation of genetic mutants in this regulatory system should allow significat experiments to be performed in the area of gene regulation.

ULTRAVIOLET MUTAGENESIS IN YEAST - RADIATION STIMULATED RECOMBINATION IN YEAST, C.W. LAWRENCE, Univ. of Rochester, School of Medicine, Rochester, New York 14620

The cytochrome c gene-protein system developed by Dr. Shermaf, in which the DNA alterations responsible for mutations can be defined, is being used to examine UV mutagenesis. UV induced reversion of a

nonsense mutant has been shown to occur by
only one of the six possible single base
pair substitutions in a UV resistant strain,
though all six can be found with other
mutagens. The basis of this specificity has
been examined by reverting the nonsense
mutant in strains containing UV sensitive
mutations of different kinds.

Experiments with synchronous cultures
of meiotic cells in the green alga
Chlamydomonas have shown that recombination
can be changed by non- lethal doses of
ionizing radiation at only two short stages
during meiosis, located in preleptotene and
zygotene/pachytene. Further, intra-genic
and inter-genic recombination respond
differently. The involvement of enzymes
which repair radiation damage in radiation
stimulated intra- and inter- genic
recombination will be studied in Yeast by
making use of radiation sensitive mutants.

Results: A mutant defective in
excision-repair has been found to have
little or no effect on the specificity, but
this can be abolished in a mutant which may
be defective in recombination-repair.

STUDY OF THE BIOLOGICAL EFFECTS INDUCED BY
FOREIGN DNA IN RECIPIENT PLANT CELLS AND
TISSUES,
L. LEDOUX, Centre Detude Lenergie Nucl.,
Brussels, Belgium

The study of the biological effects
induced by foreign DNA in recipient plant
cells and tissues will be pursued in : a)
crucifers: Arabidopsis thaliana and Sinapis
alba; b) rice; c) corn; d) monocellular
algae (such as Chlamydomonas).

The experiments will involve : 1) the
use of biochemical mutants in order to
attempt correction of metabolic deficiencies

with exogenous informative molecules; 2)
attempts to introducing new genetic
information in recipient genomes either
preexisting or not, such as lysine
information from bacteria in rice and corn.
The assays will involve the immunological
detection of the foreign gene products,
their effects at the level of the
isoenzymes, etc.

THE GENETIC CONTROL OF CHLOROPLAST STRUCTURE
AND FUNCTION IN CHLAMYDOMONAS REINHARDI,
R.P. LEVINE, Harvard University, School of
Arts, Cambridge, Massachusetts 02138

This award involves continuation of the
support provided under grant GB-18666 for
Dr. Levine's studies on the genetic control
of chloroplast structure and function in
Chlamydomonas. This research is focused on
the extent to which the structure and the
function of an organelle are determined by
genetic information contained organelle DNA
and nuclear DNA, and the degree to which
organelle ribosomes and cytoplasmic
ribosomes participate in the translation of
this information. The significance of this
research relates to the fact that the
biology of a eukaryotic cell is bound to the
genetics, structure and function of its
major cell organelles such as chloroplasts
and mitochondria. These organelles and the
cells in which they are found follow
independent courses in certain directions,
but are totally interdependent in others.
The unicellular green alga Chlamydomonas has
been chosen to investigate the nature of
this interdependence because it possesses
many favorable properties for the study of
the relationships between an organelle, the
chloroplast, and the cell in which it
occurs, and because there is a growing body

of knowledge relating to its genetics, fine structure, physiology, and biochemistry.

CELL DIFFERENTIATION DURING GAMETOGENESIS, R.P. LEVINE, Harvard University, School of Arts, Cambridge, Massachusetts 02138

Gametic differentiation will be studied in Chlamydomonas reinhardi. This includes (1) a study of the metabolic changes that are first elicited by nitrogen starvation in the two mating types; (2) a study of possible changes in nucleic acid biosynthesis and nuclear proteins during gametogenesis; (3) a study of the transcriptional and translational requirements for gametogenesis; and (4) a study of the nature of the mating-type substances and their interaction at the time of mating.

BIOCHEMICAL GENETICS OF MICROTUBULES, D.J. LUCK, Rockefeller University, Graduate School, New York, New York 10021

Conditional (temperature sensitive) mutants affecting flagellar function in Chlamydomonas reinhardi have been obtained. Six of these have been selected for intensive study of their microtubule proteins. When the temperature is shifted to non-permissive levels, all show immediate defects in flagellar function, some show fragility and loss of flagella, some show altered reaction with colchicine, and one shows additional defects in cell division. It is proposed to complete the biological and genetic characterization of these strains, and the chemical characterization

of their flagellar and cellular microtubular
proteins. Additional mutants of this type
will be obtained and further attempts to
obtain colchicine resistant mutants will be
made. A second group of mutants which show
defects in flagellar function only after a
division cycle at non-permissive
temperatures will be used to study flagellar
assembly and possible non-chromosomal
determinants of basal body function.

HORMONES CONTROLLING REPRODUCTION IN
ALLOMYCES AND OEDOGONIUM,
L. MACHLIS, Univ. of California, School of
Letters, Berkeley, California 94720

It is proposed to study the mechanism
of action of the sperm- attractant sirenin.
This will be done mostly through the use of
labeled sirenin but also by observing the
effects of various metabolic inhibitors on
the action of sirenin. The possibility of
there being species - specific sirenins will
also be studied.
It is also proposed to produce,
isolate, characterize and establish the
structure of the sperm-attractant from
Oedogonium cardiacum. Lastly, it is proposed
to attempt to establish the nutrient
requirements for the growth of Oedogonium
borisianum so as to initiate studies on the
mechanism of action of some of the hormones
already shown to control reproduction in
this organism.

PLASTID PROTEIN SYNTHESIS,
M.M. MARGULIES, Smithsonian Institution,
Rockville, Maryland 20852

Sites of synthesis of chloroplast
protein: Experiments are being carried out
to determine which proteins chloroplasts can
synthesize. Use is being made of antibiotics
such as chloramphenicol and spectinomycin
which are reported to specifically inhibit
chloroplast but not cytoplasm protein
synthesis, and of cycloheximide which is
reported to specifically inhibit cytoplasm
but not chloroplast protein synthesis. The
specificity of these antibiotics is being
tested by examining their effect on the
formation of labeled nascent protein
attached to 70 S chloroplast and 80 S
cytoplasm ribosomes in vivo using
Chlamydomonas. This work will probably be
extended to beans or spinach. The effect of
the antibiotics on incorporation of amino
acids into various chloroplast proteins is
being studied in order to determine whether
or not they are synthesized by the
chloroplast. It is one of our principal
aims to determine whether or not ribulose
diphosphate carboxylase is synthesized only
in the chloroplast. This aim is also being
investigated by studying whether or not bean
chloroplasts can incorporate amino acids
into this enzyme in vitro.
Control of plastid protein synthesis in
flowering plants: Light stimulates protein
synthesis in chloroplasts of flowering
plants. We have found that chloroplasts
from greening leaves have increased ability
to incorporate labeled amino acids into
protein compared with leaves of etiolated
plants. The basis of this observation will
be studied by examining and comparing levels
of t-RNA and amino acyl adenylate
synthetases (and other components of the
proteins synthesizing system) in etioplasts

and chloroplasts from greening leaves.

THE ODOR OF SEAWEED,
R.E. MOORE, Univ. of Hawaii, School of Arts,
Honolulu, Hawaii 96822

Isolation, structure determination, and
biosynthesis of hydrocarbons found in the
essential oil or algae of the genus
Dictyopteris and an investigation of the
odoriferous constituents of Chrondrococcus
and several phytoflagellates found in
Hawaii.

CONTROL OF REPLICATION OF CYTOPLASMIC DNAS
IN CHLAMYDOMONAS,
R. SAGER, City University of New York,
Graduate School, New York, New York 10021

The specific aim of this research is to
investigate the control of replication of
organelle DNAs. Chlamydomonas provides a
unique opportunity to study this problem
both because nuclear and chloroplast DNAs
can be readily separated in cesium chloride
equilibrium density gradients and because of
our extensive studies of cytoplasmic
genetics with this organism. We now plan to
use both genetic and biophysical approaches
to correlate the physical and genetic
behavior of organelle DNA. Replication of
chloroplast DNA is being studied by
following the incorporation of radioisotopic
adenine of $15N-NH_4$ ion into total cell DNA,
which is then extracted and fractionated in
cesium chloride density gradients. We are
examining the time and extent of replication
of chloroplast DNA in cells grown under
different physiological regimes, and

examining the ability of various growth inhibitors to uncouple the replication of chloroplast and of nuclear DNAs. Detailed studies of the fate of the chloroplast DNAs from male and female parents, correlating physical and genetic data have been in progress for several years and are continuing.

REGULATION OF ENZYME SYNTHESIS AND DEGRADATION, AND GENE EXPRESSION DURING THE PLANT CELL CYCLE,
R.R. SCHMIDT, Virginia Polytechnic Institute, School of Agriculture, Blacksburg, Virginia 24061

OBJECTIVE: Study the biochemical mechanisms regulating the timing of expression of structural genes coding for enzymes catalyzing steps in nutrient assimilation (i.e. carbon dioxide fixation, nitrate and nitrite reduction, assimilation of exogenous ammonium and urea) and in ribonucleotide and deoxyribonucleotide biosynthesis, and to determine the mechanisms regulating the synthesis and degradation of these enzymes during the cell cycles of lower and higher eucaryotic plant cells. The intracellular location and site of synthesis of these enzymes and their genes will also be elucidated.
APPROACH: Synchronous cultures of both lower and higher plant cells in suspension culture will be prepared by a new isopycnic density- gradient technique. The biochemistry of the synchronous cell cycle will be approached using genetic analysis, subcellular organelle isolation, differential inhibitors of organelle and cytoplasmic enzyme synthesis, specific

188

inhibitors of gene replication and expression, and enzyme fractionation and purification.

PROGRESS: Two isozymes of glutamate dehydrogenase were discovered in a thermophilic strain of Chlorella pyrenoidosa. The NADPH:NADH coenzyme activity ratios for the NADH- and NADPH-specific isozymes were 1:5 and 33:1, and the molecular weights were estimated to be 179,000 and 269,000, respectively. Only the NADH-specific isozyme was detectable in nitrate-cultured cells; the synthesis of the NADPH-specific isozyme was inducible by ammonium. By use of actinomycin D and cycloheximide as selective inhibitors, the induction of the NADPH-specific isozyme was shown to be dependent upon both RNA and protein synthesis, respectively. This isozyme was inducible at all times during the cell cycle. The potential, i.e., maximum rate of induction, remained constant and then increased in a single step during the period of DNA replication, and the fold-increase in the potential and in DNA was essentially equal. These data indicate that the structural gene for this isozyme is continuously available for transcription during the cell cycle of this eucaryote, and are consistent with hypothesis that, under fully-induced conditions, the gene dosage governs the change in potential. These data have never been obtained for any eucaryote and represent "breakthrough" information.

A ROLE FOR ORGANELLES IN DEVELOPMENT, D.C. SHEPHARD, Case Western Reserve Univ., School of Medicine, Cleveland, Ohio 44106

The objective of this research project is to systematically determine the genetic information possessed by an organelle in a

cell- organelle system that is uniquely
suited to such an investigation - the
chloroplasts of Acetabularia mediterranea.
   In order to determine whether the
syntheses of plastic RNA and protein species
occur within the organelle, biosynthetic
activities in vivo (intact cells) and in
vitro (using a chloroplast isolate) are
being compared. Then by assaying the
constituent or its rate of synthesis during
several weeks of enucleate growth, an
experiment uniquely possible with this alga,
and by studying the effects of drugs such as
rifampicin in enucleate cells it is possible
to determine whether the constituent is
coded in the chloroplast genome. Ultimately
the comparative roles of nuclear and
chloroplast genes in determining chloroplast
structure and function should become clear.
   A comparison of the plastid
biosyntheses in enucleate cells and in
isolated chloroplasts with those of normal
cells should be of value, not only by
indicating the role organelle genes could
play in cell differentiation, but also by
suggesting control mechanisms for short term
physiological adaptations involving
organelle function. It can also be stated
that if organelle genes are shown to play a
considerable role in controlling normal cell
processes then serious and heritable
abnormalities in cell function may result
from disturbances of the organelle genetic
system.

CHROMOSOME STRUCTURE AND PROTEIN SYNTHESIS,
H. SWIFT, Univ. of Chicago, School of
Medicine, Chicago, Illinois 60637

   Research of students and postdoctoral
fellows is concerned primarily with problems
related to chromosome structure and function

in eucaryote cells, and with various
cellular aspects of protein synthesis. Most
work involves a combination of biochemical,
cytochemical, and morphological techniques,
including electron microscopy,
autoradiagraphy, microphotometry and various
kinds of cell fractionation and
centrifugation. Problems under study
include work on specific histone fractions
from isolated nuclei of cells in various
synthetic states, the relation between
redundant DNA sequences and AT polymers to
chromosome structure, chromosomal structure
in relation to processes of gene
amplification, denaturation-mapping of
mitochondrial DNA, mechanisms influencing
mitochondrial number and DNA turnover, and
studies on chloroplast DNA and chloroplast
structure during zygote formation in
Chlamydomonas.

BIOSYNTHESIS OF AMINO ACIDS, PEPTIDES, AND
PROTEINS,
J.F. THOMPSON, State University of New York,
U.S.D.A. Soil & Wa. Cons. Div., Ithaca, New
York 14850

OBJECTIVE: Determine the biochemical
pathways of formation and metabolism of
amino acids, proteins, and related
substances in plants and animals.
APPROACH: The effects of mineral
nutrition and environmental factors upon the
amino acid and protein components of plants
are determined. Pathways of synthesis and
metabolism of amino acids and proteins are
studied in living and in cell-free systems.
The structures and functions of RNA in
protein synthesis are determined.
Mechanisms controlling the synthesis of
amino acids, proteins, and related
substances are identified and characterized.

PROGRESS: The determination of the structure of the transfer ribonucleic acid incorporating lysine has been nearly completed. One unidentified nucleotide appears to be a new "rare nucleotide" and its instability has hampered its identification. In the utilization of urea by Chlorella, two enzymes are involved. One carboxylates urea to allophanate and requires ATP, Mg, CO(2), and biotin. The other hydrolyzes allophanate to NH(4) and CO(2). Higher plants contain an enzyme which hydrolyzes urea directly without the ATP requiring step found in Chlorella. Proline accumulates in plants under water stress. Acetylation of glutamic acid in plants takes place with the acetyl donor being either acetyl ornithine or acetyl CoA. It appears that two enzymes are involved and both are activated by NH(4). High concentration of aconitic acid in grasses are frequent during outbreaks of grass tetany. Leaves, but not roots, of wheat plants may have ten times as much aconitic acid when grown at high-K levels as when grown at low-K levels.

CHLOROPLAST ENDOSYMBIOSIS AND BIOCHEMICAL AUTONOMY,
R.K. TRENCH, Yale University, School of Arts, New Haven, Connecticut 06520

This research concerns the investigation of aspects of chloroplast endosymbiosis and biochemical autonomy. The unique association of chloroplasts from certain marine algae (Siphonales) which occur naturally as functional endosymbionts in certain gastropod molluscs (Sacoglossa) will be studied. These chloroplasts can

continue to function in the animal cell
environment for as much as six weeks.

Since it is unlikely that the animal
cell nucleus codes information which directs
synthesis of macromolecules in chloroplasts,
any DNA- dependent macromolecular synthesis
which can be demonstrated by symbiotic
chloroplasts in situ would have to be
directed by chloroplast DNA (the plastome).

In continuing studies of chloroplast
endosymbiosis, the extent to which
chloroplasts can direct synthesis of
structural and enzyme proteins; whether
symbiotic chloroplasts can synthesize
chloroplast specific RNA and DNA; the
mechanism of phagocytosis of chloroplasts by
the slug's digestive cells (i.e. the cells
which sequester and maintain the
chloroplasts), and the in vitro
physiological and biochemical
characteristics of chloroplasts of
Siphonales will be studied in order to gain
further insight into their predilection for
establishing endosymbiotic associations.

ANTI-ALGAE COMPOUNDS,
R.T. VANALLER, Univ. of Southern Mississippi,
School of Science, Hattiesburg, Mississippi
39401

Isolation of and structural studies on
a naturally occurring inhibitor of algae.
The material inhibits growth of green and
blue- green algae commonly found in
pollution.

BIOCHEMICAL EFFECT OF GLYOXALASE IN ALGAE,
B.D. VANCE, North Texas State University,
School of Arts, Denton, Texas 76203

Previous work concerned with the biochemical effects of glyoxalase in green algae showed that cell division was enhanced and suggested that levels of indole-3-acetic acid (IAA) were affected.

Endogenous levels of IAA were measured in synchronous cultures of Chlorella pyrenoidosa. The cultures were synchronized by alternating light: dark periods of 15:9 hr at a temperature of 40 plus or minus 1 C. After synchrony was attained cultures were exposed to a low light treatment 350 ft-C. The time to incipient cell division was 6 hr and 15 min. Samples were taken at 3 sampling periods during the low light treatment period. Algal extracts were analyzed by a fluorometric procedure which measured the indole-gamma-pyrone product formed by the action of the trifluoracetic acid-acetic anhydride reagent on IAA. IAA levels increased gradually from the autospore stage to the adolescent stage and more rapidly when approaching the ripened adult stage. The total percentage increase of IAA from autospore to adult was 180:3 per cent. Levels of IAA were two times higher just prior to division than in the autospore stage.